International Conference on Technology and Vocational Teachers (ICTVT-2018)

IOP Conference Series: Materials Science and Engineering
Volume 535

Yogyakarta, Indonesia
15 November 2018

Editors:

Suprapto
Didik Nurhadiyanto
Moh. Khairudin
Istanto Wahyu Djatmiko
Putu Sudira
Bernadus Sentot Wijanarka

ISBN: 978-1-5108-8921-7
ISSN: 1757-8981

Printed from e-media with permission by:

Curran Associates, Inc.
57 Morehouse Lane
Red Hook, NY 12571

Some format issues inherent in the e-media version may also appear in this print version.

No changes have been made to the content of these proceedings. There may be changes to pagination and minor adjustments for aesthetics.

Printed by Curran Associates, Inc. (2019)

For permission requests, please contact the Institute of Physics at the address below.

Institute of Physics
Dirac House, Temple Back
Bristol BS1 6BE UK

Phone: 44 1 17 929 7481
Fax: 44 1 17 920 0979

techtracking@iop.org

Additional copies of this publication are available from:

Curran Associates, Inc.
57 Morehouse Lane
Red Hook, NY 12571 USA
Phone: 845-758-0400
Fax: 845-758-2633
Email: curran@proceedings.com
Web: www.proceedings.com

Table of contents

Volume 535

International Conference on Technology and Vocational Teachers (ICTVT-2018)

15 November 2018, Yogyakarta, Indonesia

Accepted papers received: 16 April 2019

Preface

Papers

Engineering and Technology

IOP Conf. Series: Materials Science and Engineering **535** (2019) 011001 doi:10.1088/1757-899X/535/1/011001

ICTVT

INTERNATIONAL CONFERENCE ON TECHNOLOGY AND VOCATIONAL TEACHERS

The 2nd ICTVT 2018:

Innovation and Development of Technology for Vocational Education

ICTVT
International Conference
on Technology and Vocational Teachers

Organized by:
Faculty of Engineering
Yogyakarta State University, Indonesia.
November 15th, 2018

IOP Conf. Series: Materials Science and Engineering **535** (2019) 011001 doi:10.1088/1757-899X/535/1/011001

Preface

I warmly welcome you to The Second International Conference on Technology and Vocational Teachers (ICTVT), Yogyakarta, 15 November 2018. It's a great pleasure on behalf of ICTVT commitee to have you today to participate and to share your studies, ideas, knowledge and experience in this conference. Through ICTVT 2018, we would like to engage with all of you in an open and constructive dialogue about **Innovation and Development of Technology for Vocational Education.**

The recent trend of automation and data exchange in manufacturing technologies, or Industry 4.0, assists companies to identify and implement Industry 4.0 scenarios, such as Interoperability, Information Transparency, Technical Assistance, and Decentralized Decisions.

This new industry era needs some development and innovation for several related fields. Innovation and development of industry 4.0 stimulate and demand the development of education, especially vocational education. Therefore, this conference aims to explore the role and potential of technology for vocational education as the backbone of the industry era 4.0.

We are honored to have outstanding speakers with us today. All three are esteemed specialists in the field. We are grateful for the time and effort they took to share their works and experiences.

There are 67 papers have been submitted at the 2th ICTVT but only 38 of these have been accepted for the presentation after a blind peer review process and will be presented at 5 parallel oral sessions. I sincerely thank the program commitee members for their thorough and timely reviews of the papers.

Last but not least, I would like to thank the organizing committee who work so hard to prepare all aspects of the conference. The success of this conference depends on their commitment, dedication and generous contribution.

It has been a great privilege for me to serve as the General Chair of ICTVT 2018 and it is my hope that you find the conference stimulating, fulfilling and enjoyable. I thank you for your support and your attendance, and wish you a pleasant experience in ICTVT 2018.

Salam

Moh. Adam Jerusalem, Ph.D.

Conference Chair of ICTVT

IOP Conf. Series: Materials Science and Engineering **535** (2019) 011001 doi:10.1088/1757-899X/535/1/011001

Words from the Rektor

Assalamu'alaikum Warahmatullah Wabarakatuh.

May peace and god's blessings be upon all of us.

Entering the industrial revolution 4.0 era, higher education is demanded to prepare about one hundred and thirteen million skilled workforces. This demand has led the Indonesian government to promote the five priority industries to compete the future of work in 2030, including food and beverage, textile & apparel, automotive, electronic, and chemical industries. All those five priorities require strong support in the preparation of qualified workforce from relevant educational institutions, particularly vocational education institutions in higher education.

As one of the pioneer universities of vocational education in Indonesia, UNY has committed to dedicate more efforts and energy to help foster the development of education, especially vocational education to face the industrial revolution 4.0 era. To give more contributions, UNY hosts *The* ***International Conference on Technology and Vocational Teachers (ICTVT) 2018,*** as a forum to share research findings, best practices, and ideas in regard to **innovation and development of technology for vocational education.**

By having the theme of "Innovation and Development of Technology for Vocational Education", it is expected that this forum enlightens vocational teachers and practitioners in facing the challenge of utilizing and integrating technologies into their instructional process as well as designing useful learning environments in the context of real world activities for vocational students. This forum is also projected to encourage researchers, vocational teachers and students to present and publish their research findings further in international reputable journals.

Finally, my appreciation and gratitude are for all committee members and those who have been working hard in preparing all important aspects of this conference. Our special thanks also go to all the invited speakers and parallel speakers who have been willing to share ideas in ICTVT 2018.

Wassalamu'alaikum Warrahmatullah Wabarakatuh.

Yogyakarta, 6 November 2018
Rector of Universitas Negeri Yogyakarta
Prof. Dr. Sutrisna Wibawa, M.Pd.

IOP Conf. Series: Materials Science and Engineering **535** (2019) 011001 doi:10.1088/1757-899X/535/1/011001

Peer Review General Policies

Papers will be rejected as refereed paper if a sub-mission deadline is missed or if maximum capacity has been reached. Paper will then publish as an unrefereed paper at the ICTVT website (http://ictvt2018.uny.ac.id), we only received is allowed in general.

1. Paper will reject if submission or correction deadline is missed
2. Each paper shall have two reviewers
3. A reviewer shall not know the identity of another reviewer assigned to the paper
4. A reviewer shall not contact the author directly concerning the paper
5. Reviewer shall not make any personal remarks or comments that may betray their identity when entering the instructions for requested correction/revision of the paper
6. Reviewer must not referee papers of which they are author or co-authors
7. There are only 3 possible outcomes from reviewers in first cycle:
   a. Excellent
   b. Solid
   c. Accepted if possible
   d. Rejected if possible
   e. Reject
8. There are 2 possible outcomes from reviewers in second cycle:
   a. Accepted
   b. Rejected
9. Decisions about paper rejection are final

Peer Review Acceptance Criteria

All papers published in this volume of IOP Conference Series: Materials Science and Engineering have been peer reviewed through processes administered by the proceedings Editors. Reviews were conducted by expert referees to the professional and scientific standards expected of a proceedings journal published by IOP Publishing. Here are the criteria:

1. The quality of content must have solid work and decent contribution
2. The paper must at least have significant for the theory and practices
3. Paper must discuss one step forward from the ordinary discussion
4. The paper should comply the theme that has been given
5. The paper quality has to be well written

Question posed to referee in review module

Reviewer must acknowledge they have read the referee guidance statements. Reviewer answer Yes/No to the following queries

1. Is the quality of content meet the requirement?
2. Is the significance for theory or practices meet the requirement?
3. Is there any innovative aspect in the paper?
4. Is the paper's theme align with the conference theme?
5. Is the paper present in proper way?

Outcome of the Referee Dialogue

1. If answer to 1) 2) and 3) are negative (no) then paper is uncorrectable (automatic rejection)
2. If answer to 1) – 5) are positive (yes) then paper is automatic accepted
3. If answer to 1) – 3) are positive but either 4) or 5) is negative then two cases exist
   a. Easy/straight forward to correct then: revise
   b. Difficult to correct then: reject

Reviewer then enters final appraisal

Question: Is the paper acceptable for publication in its present form?
Answer is limited to three possibilities (entered by the reviewer)

1. Yes (accept)
2. No, but easily corrected (request author to make revision)
3. No and uncorrectable (reject)

IOP Conf. Series: Materials Science and Engineering **535** (2019) 011001 doi:10.1088/1757-899X/535/1/011001

List of Comitee

Patron

Prof. Dr. Sutrisna Wibawa, M.Pd. (Rector of Universitas Negeri Yogyakarta)

Steering Commitee

Dr. Widarto, M.Pd. (Dean of Faculty of Engineering)
Agus Santoso M.Pd. (Vice Dean of Faculty of Engineering)
Dr. Giri Wiyono, M.T. (Vice Dean of Faculty of Engineering)

Program Committee Member (Reviewer)

1. *Dr. Noraziah Ahmad, Universiti Malaysia Pahang (Scopus ID: 22939129100)*
2. *Dr. Wan Maseri Wan Mohd, Universiti Malaysia Pahang (Scopus ID: 43061604400)*
3. *Dr. Zailani Abdullah, Universiti Malaysia Terengganu (Scopus ID: 36170180000)*
4. *Dr. Abdullah Khan, Agriculture University Peshawar, Pakistan (Scopus ID: 55803876600)*
5. *Dr. Rabiei Mamat, Universiti Malaysia Terengganu (Scopus ID: 55899244600)*
6. *Dr. Shamsudeen Sanni, Kampala international University, Uganda (Scopus ID: 36680505300)*
7. *Dr. Lubna A Gabralla, Sudan University of Science Technology (Scopus ID: 55954959400)*
8. *Dr. Safii Muhammad Abdulhamid, Federal University of Technology Minna, Nigeria (Scopus ID: 56157617800)*
9. *Dr. Nader Sohrabi Safa, Warwick University UK (Scopus ID: 55683063600)*
10. *Dr. Mohd Isa Awang, UNISZA Malaysia (Scopus ID 54884552300)*
11. *Dr. M Zubair Rehman, UTHM Malaysia (Scopus ID: 55574123132)*
12. *Dr. Abul Hasheem Beg, Charles Sturt University, Australia (Scopus ID: 36056069000)*
13. *Dr. Adamu I Abubakar, IIUM Malaysia (Scopus ID: 54891756900)*
14. *Dr. Ahmad Nazari Mohd Rose, UNISZA Malaysia (Scopus ID: 26667997300)*
15. *Dr. Abdulsalam Ya'u Gital, Abubakar Tafawa Balewa University, Bauchi, Nigeria (Scopus ID: 56077965400)*
16. *Dr. Ahmad Shukri Mohd Noor, Universiti Malaysia Terengganu (Scopus ID: 35198623500)*
17. *Dr. Habib Shah, Islamic University of Madinah, Saudi Arabia (Scopus ID: 55077194100)*
18. *Dr. Younes Saadi, Batna University, Algeria (Scopus ID: 55602551000)*
19. *Dr. Haruna Chiroma, Federal College of Education, Nigeria (Scopus ID: 55583663400)*
20. *Prof. Koichi Goda, Ph.D. (Yamaguchi University, Japan)*
21. *Prof. Dr. Jailani Bin Md Yunos (University Tun Hussein On Malaysia)*
22. *Dr. Margarita Pavlova, (Scopus ID: 11838861100), The Education University of Hong Kong*
23. *Dr. Maizam Binti Alias, (Scopus ID: 8344647400), University Tun Hussein Onn Malaysia*
24. *Ming Chang Wu, Ph.D., National Yunlin University of Science and Technology, Taiwan.*
25. *Prof. Herman Dwi Surjono, Drs., M.Sc., Mt., Ph.D. (Universitas Negeri Yogyakarta)*
26. *Assoc. Prof. Dr. Mochamad Bruri Triyono, M.Pd. (Universitas Negeri Yogyakarta)*
27. Assoc. Prof. Tutut Herawan, Ph.D. (University of Malaya, Malaysia)
28. *Assoc. Prof. Dr. Zaharuddin Mohamed (University Technology Malaysia)*
29. *Dr. Siti Hamidah (Universitas Negeri Yogyakarta)*

IOP Conf. Series: Materials Science and Engineering **535** (2019) 011001 doi:10.1088/1757-899X/535/1/011001

Organizing Committee:

1. M. Adam Jerusalem, Ph.D (Chairperson, Scopus ID: 57191591833)
2. Dr. Ratna Wardani
3. Pipit Utami, M.Pd.
4. Saila MA, S.Pd.
5. Hening Harjanti, S.Pd.
6. Faranita Surwi, M.T.
7. Tri Siswanti, S.E.
8. Nur Hidayat, M.Pd.
9. Achmad Arifin, M.Eng.
10. Surono, M.Pd.
11. Dr. Badraningsih Lastariwati,
12. Dewi Eka Murniati, M.M.
13. Muhfizaturrahmah, M.T.
14. Sigit Pambudi, M.T.
15. Andik Asmara, M.Pd.
16. Haryo Aji, S.Pd.

Proceedings Editor:

1. Suprapto, Ph.D. , (Scopus ID: 56580451100), Universitas Negeri Yogyakarta.
2. Dr. Eng. Didik Nurhadiyanto, M.T., (Scopus ID: 55787020500), Universitas Negeri Yogyakarta.
3. Moh. Khairudin, Ph.D., (Scopus ID: 37361115800), Universitas Negeri Yogyakarta.
4. Dr. Istanto Wahyu Djatmiko, Universitas Negeri Yogyakarta.
5. Dr. Putu Sudira, Universitas Negeri Yogyakarta.
6. Dr. Bernadus Sentot Wijanarka, Universitas Negeri Yogyakarta.

IOP Conf. Series: Materials Science and Engineering **535** (2019) 011002 doi:10.1088/1757-899X/535/1/011002

Peer review statement

All papers published in this volume of *IOP Conference Series: Materials Science and Engineering* have been peer reviewed through processes administered by the proceedings Editors. Reviews were conducted by expert referees to the professional and scientific standards expected of a proceedings journal published by IOP Publishing.

IOP Conf. Series: Materials Science and Engineering 535 (2019) 012001 doi:10.1088/1757-899X/535/1/012001

Analysis of bamboo mechanical properties as construction eco-friendly materials to minimizing global warming effect

I Yasin[1] and A Priyanto[2]

[1,2]Departement of Civil Engineering, Sarjanawiyata Tamansiswa University, Yogyakarta, Indonesia

E-mail: iskandartamansiswa@gmail.com

Abstract. Burning fossil fuels increases the amount of carbon dioxide in the atmosphere. Carbon dioxide is the main cause of the greenhouse effect and global warming. Climate change causes an increase in the number of human deaths due to diseases and natural disasters. In general, in the past decade there has been deforestation of 13 million ha / year. The highest deforestation rates (> 3.4 million ha / year) occur in South America and Africa while in Asia also experience deforestation in many places. Bamboo forests are very important especially in East and Southeast Asia as well as in Africa. The potential of bamboo is very large in controlling soil erosion, water conservation, land rehabilitation and carbon sequestration. In carbon sequestration bamboo contributes very significantly. The use of bamboo as an environmentally friendly construction material is very important to do. Testing of bamboo mechanical properties based on ISO 3129-1975 standards, which includes testing of fiber parallel compressive strength, parallel fiber shear strength, fiber parallel tensile strength and straight fiber tagak flexural strength. Setting-up testing is carried out after each specimen in the condition of the kiln dry water content is in the range 12-13% according to the requirements of the ISO 3129-1975 standard. Analysis of variants (Anova) of pressure treatment on bamboo material has an effect on some of the mechanical properties of bamboo. Pressure treatments up to 2.5 MPa have a significant effect on the elastic modulus, tensile strength and shear strength of bamboo while the compressive strength is not significantly affected by the pressure treatment of bamboo material. Bamboo forests have a big influence in reducing the impact of global warming and climate change.

1. Introduction

As a result of the use of fossil fuels and nuclear power, the temperature of the heat generated flows into the atmosphere. Since 1880 global warming has accumulated in air, water and land. In general, thermal pollution contributes significantly to 55% of global warming [1]. In the last 100 years, the surface of the earth has risen to an average temperature of 1 degree Celsius or 1.3 degrees Fahrenheit. Burning fossil fuels increases the amount of carbon dioxide in the atmosphere. Carbon dioxide is the main cause of the greenhouse effect and global warming. Climate change causes an increase in the number of human deaths due to diseases and natural disasters [2]. From the June 1972 environmental conference in Stockholm to the June 1992 Earth Summit in Rio de Janeiro and other international conferences on the environment, polemic for environmental sustainability increased significantly and decided on international actions to reduce carbon dioxide and global warming in earth. There are three carbon stocks on earth, namely the oceans, the atmosphere and terrestrial ecosystems. Forests are able to store the largest carbon in terrestrial ecosystems, namely 1146 x 1015 g carbon or 56% of total carbon in terrestrial ecosystems [3]. So it needs fast-growing plants and is able to absorb carbon in large capacity to restore forest sustainability.

In general, in the past decade there has been deforestation of 13 million ha / year. The highest deforestation rates (> 3.4 million ha / year) occur in South America and Africa while in Asia also experience deforestation in many places. Deforestation dramatically affects the human population that lives from forests. The survival of indigenous peoples and traditional communities is threatened by deforestation. Changes in rainfall include the impact of deforestation which causes the risk of forest fires in the dry season. In addition, it also causes negative changes in water quality and biodiversity losses because more than half of the animals that live on the earth depend on forests for their survival. In general the causes of deforestation are infrastructure development, expansion of agricultural and plantation land, settlement and mining expansion [4].

Bamboo forest ecosystems are an important part of the forest ecosystem and carbon sink on the earth [3]. Bamboo is a renewable natural resource that has the highest growth speed [5]. As one of the fastest growing plants, bamboo growth is 30 mm to 100 mm per day during growth. The height of bamboo can reach 36 meters with a diameter of 1-30 cm [6]. Bamboo has more than 70 genera and more than 1000 species in botanical literature. Bamboo plants can adapt to a short life cycle. Bamboo growth can reach 70 mm per day and can reach 350 - 450 mm [7]. The development of the stem is completed in 4-6 months and bamboo reaches adults only takes 2 - 6 years, depending on the species [8]. Cylindrical bamboo stems with a diameter of 29-300 mm, 60-70% of the stem consists of fiber. The average height of bamboo is about 100 times its diameter [9]. In nature bamboo is the fastest growing plant [10].

Bamboo forests are very important especially in East and Southeast Asia as well as in Africa. The potential of bamboo is very large in controlling soil erosion, water conservation, land rehabilitation and carbon sequestration [11]. One hectare of bamboo plants can absorb more than 12 tons of carbon dioxide in free air. With the preservation of bamboo means having a carbon dioxide suction machine in large capacity [12]. In the *Bambusoideae* family, bamboo is a giant grass with a shell shape, orthotropic, has a high strength in the longitudinal direction whereas in the transverse direction it has a low strength. Bamboo fibers are denser on the outer skin for resistance to wind gusts during their growing period [13].

Fig. 1. Bamboo Fiber Structure of *Dendrocalamus asper*.

2. Literatur Review

In carbon sequestration bamboo contributes very significantly. About 80% of the area containing bamboo is in the tropical regions of South and Southeast Asia [11]. The ability of bamboo in carbon sequestration tends to be second to none compared to other plants because of the speed of biomass accumulation and the effective fixation of solar energy and carbon dioxide. A quarter of biomass in the tropics and one fifth in the subtropical region comes from bamboo. The development of bamboo can reach 15 km from the center of growth with a stem diameter reaching 30 cm [14]. Bamboo clumps can hold up to 25% of rain water falling to the ground, this percentage is far greater than conifers and pine plants. Bamboo is able to withstand large amounts of rainfall due to the thick layer of leaves and strong clumps. Interception of bamboo clumps depends on stem quantity and leaf area index [15]. The annual interception of rainfall in the *Dendrocalamopsis latiflorus* bamboo forest is 128.1 mm with a medium distance of 3 x 4 meters and an overall density of 5000 bamboo stems per hectare. The interception ratio is 14.5% with runoff coefficient of 7.47%. So that bamboo clumps have a positive correlation between clump density and rainfall [11]. Bamboo leaf drop effectively keeps the soil moist and reduces water evaporation from the ground. Bamboo leaf waste can hold 2.75 times more weight than its own dry weight. Bamboo leaves have high water retention capacity and can increase soil organic content [16]. The capacity to hold leaf waste water in bamboo forests with a mixture of species of *Phyllostachys pubescens* and *Cunninghamia lanceolata* reached 21.29 $t.hm^{-2}$, much higher if bamboo forests which only consisted of *Cunninghamia lanceolata* reached 7.37 $t.hm^{-2}$. *Dendrocalamus latiflorus* bamboo leaf waste with a density of 825 stems per hm^2 has a capacity of absorbing moisture from 2.7 to 2.9 times the dry weight [17]. The same conclusion regarding the capacity of bamboo leaves in moisture absorption has been produced by other researchers ([17], [18], [19], and [20]).

Erosion is one of the most severe types of soil degradation that threatens land productivity and causes loss of topsoil. Bamboo is one of the most effective in controlling erosion. Bamboo which is commonly used for erosion control is *Bambusa vulgaris* [14]; [21]. The advantages of bamboo are the extensive fabrication of root fibers, interconnected rhizomes and always produce new stems that allow harvest without damaging the soil, relatively dense leaf lush which protects soil from high rainfall [11]. For erosion control purposes, in Japan bamboo is planted in places that are prone to erosion. In southwest Japan, bamboo is planted in coastal areas overlooking the Pacific Ocean. Especially in Kagoshima prefecture bamboo functions as a protector of hilly erosion called "Karami" for erosion protection for more than 100 years [22]. *Sasa* and *Indocalamus* bamboo clumps are spread in the Hakoneyama mountainous region of Japan at an altitude of 1,000 meters above sea level. *Bambusa blumeana* and *Phyllostachys pubescens* in Brazil are used for erosion control, preventing loss of soil nutrients and improving soil structure [23]. About 90% of bamboo forests in China grow naturally upstream of rivers, lakes and along the banks of riverbanks. Of the total forest area, the contribution of bamboo forests in the Yangtse river basin is 5%, the Pear river valley is 4.5% and the Huaihe river valley is 2.5% [24]. On the Dayingjiang river in Yunnan province and on the Jiulongjian river in Fujian province, bamboo plants have succeeded in protecting riverbanks after tree planting efforts have failed to protect riverbanks. Bamboo forests are very important in the protection of riverbanks [14]; [25].

Most of the bamboo rhizomes and roots are at 0 - 30 cm from the ground so that the roots and rhizomes can function best in controlling soil erosion. Roots and rhizomes are shaped like woven mats so they are effective in protecting the soil. In this case the bamboo species that has been studied is *Phyllostachys pubescens* [26]; [27]; [28]; [29], *Pseudosasa amabilis* [30], *Phyllostachys praecox* [31]; [32], *Qiongzhuea tunfdoda* [7], *Phyllostachys makinoi* [15], *Phyllostachys bisetti* [40], *Dendrocalamus latiflorus* [33], *Dendrocalamus oldhami* [34] and *Bambusa tulda* [21]. Monopodial bamboo with creeping rhizomes, total length of rhizomes per 1 hectare of bamboo *Phyllostachys heterocycla, Phyllostachys viridis* and *Phyllostachys nigra* each from 50 - 170 km, 90 - 250 km and 200 - 320 km [35]. Sympodial bamboo root Bambusa tulda with short rhizomes can reach horizontal areas up to a distance of 5,2 meters. Roots and rhizomes though grow superficially but are horizontally widened [21].

In the past five decades, 1.2 billion hm^2 of land (11% of total earth vegetation) has been degraded and damaged by its original biotic function [36]. A quarter of land degradation is caused by humans (agricultural practices, excessive livestock grazing, deforestation, etc.) [37]. With dense leaf clumps, large amounts of biomass accumulation and abundant literfall, bamboo plays a major role in efforts to rehabilitate degraded land. In China, India and Thailand, bamboo agro-forestry models for degraded land have been developed [38].

The use of bamboo as an environmentally friendly construction material is very important to do. Stems in a bamboo clump at the age of 5-6 years can be harvested continuously without the need to plant again because the growth of bamboo shoots is fast growing. The rapid growth of bamboo is able to reduce the problems resulting from global warming on earth. Bamboo is a renewable natural resource that is the fastest growing and the fastest absorbing carbon in the air.

3. Problem Identification

At this time the timber with good quality has been difficult to obtain, so the wood is increasingly rare for buildingconstruction. To reduce deforestation due to logging for construction materials, it is time to replace wood with bamboo. Bamboo has a good strength when used properly for construction [39]. The tensile strength of bamboo parallel fibers without the nodia of the *Bambusa arundinaceace* skin much larger than the tensile strength of the reinforcing steel, and the tensile strength of the bamboo fiber in all the thickness of *Dendrocalamus asper* [40]. Tensile strength bamboo fiber ranging between (150 - 320) MPa, greater than the tensile strength of wood fibers ranges (34 - 220) MPa [41]. Parenchyma sell structures more dense with age bamboo, the bamboo optimum density occurs in 4 years old [42]. Describes the use of bamboo which is divided into a traditional bamboo (conventional) and bamboo engineering (experience the manufacturing process) [43].

Bamboo as a wood substitute material that is environmentally friendly, strength and stiffness needs further research. Abundant supply of bamboo is abundant, it is very effective and efficient because it doesn't require many personnel and conventional method used [38]. Bamboo as an environmentally friendly and renewable material, can influence the effects of global warming, maintain the sustainability of the green environment and make the earth a comfortable and friendly place for human life.

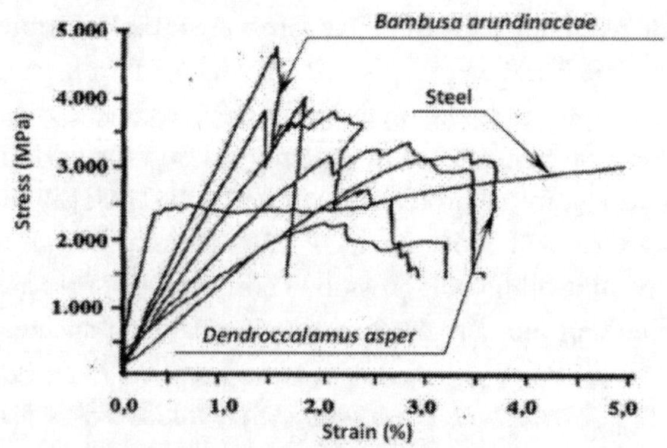

Fig 2. Bamboo Stress-Strain Graphic [40].

4. Objectives

The objective of this research to analyze the mechanical properties of bamboo so that it can know the strength and stiffness of bamboo material as a wood substitute construction material. In the construction process, bamboo material gets a pressure treatment especially on the connection, in the process of getting the desired geometric shape, in making laminated beams etc.

5. Method

Testing of bamboo mechanical properties based on ISO 3129-1975 standards, which includes testing of fiber parallel compressive strength, parallel fiber shear strength, fiber parallel tensile strength and straight fiber tagak flexural strength. Setting-up testing is carried out after each specimen in the condition of the kiln dry water content is in the range 12-13% according to the requirements of the ISO 3129-1975 standard. Adapaun specimens of specimens are as in Table 1. Testing of bamboo mechanical properties using UTM (Universal Testing Machine) and observation of changes in bamboo fiber was observed using a SEM (Scanning Electron Microscopy).

Table. 1. Mechanical Properties Testing of *Dendrocalamus asper* Specimen.

No.	Mechanical Properties of Specimen	Without Lateral Compression	With Lateral Compression		
			1,5 MPa	2 MPa	2,5 MPa
1	Compression Strength	10	10	10	10
2	Shear Strength	10	10	10	10
3	Tensile Strength	10	10	10	10
4	MoE	10	10	10	10

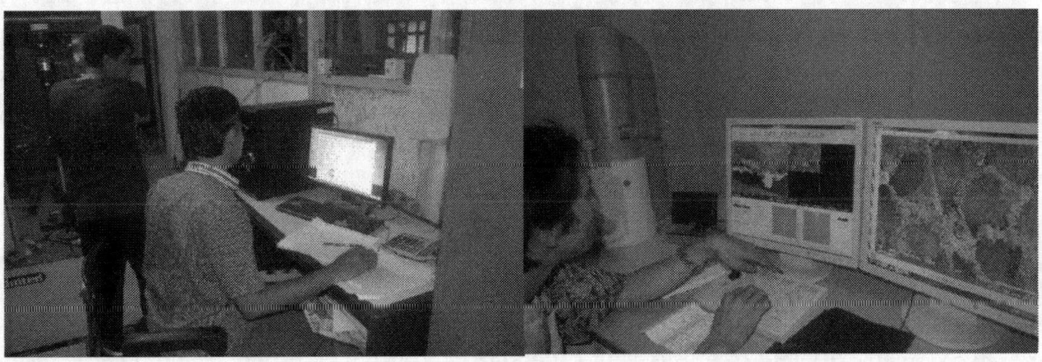

(a) (b)

Fig 3. Testing Mechanical Properties Bamboo Specimen use UTM (a) and SEM (b).

6. Result and Discussion

Tests for water content of specimens without minimal pressure were 5.47% and a maximum of 13.97% with an average of 12.95%. At a pressure of 1.5 MPa the water content is at least 6.80% and a maximum of 19.42% with an average of 12.81%. The amount of water at a pressure of 2 MPa is a minimum of 6.47% and a maximum of 16.48% with an average of 12.52%. Water content pressure of 2.5 MPa is a minimum of 8.39% and a maximum of 13.87% with an average water content of 12.43%.

The gravity of the specimen without pressure is a minimum of 0.56 gr / cm^3 and a maximum of 0.70 gr / cm^3 with an average of 0.64 gr / cm^3. At a pressure of 1.5 MPa the gravity value is a minimum of 0.66 gr / cm^3 and a maximum of 0.91 gr / cm^3 with an average of 0.73 gr / cm^3. The gravity value at a pressure of 2 MPa is a minimum of 0.62 gr / cm^3 and a maximum of 0.92 gr / cm^3 with an average value of 0.75 gr / cm^3. At a pressure of 2.5 MPa the gravity of bamboo is a minimum of 0.70 gr / cm^3 and a maximum of 0.93 gr / cm^3 with an average value of 0.78 gr / cm^3. The partial correlation of gravity with the pressure pressure variation is not significantly related. This is because pressure presses with 2.5 MPa have not damaged the structure of bamboo fibers.

The average MoE of bamboo without pressure is 152.10 MPa, the average MoE at a pressure of 1.5 MPa is 155.75 MPa, the average MoE at a pressure of 2 MPa is 157.62 MPa and the average MoE at a pressure of 2.5 MPa is 153.00 MPa. In the results of this test, the MoE of the bamboo with 2 MPa was higher than the other variations of pressing. The average MoE without pressure is 17955.77, the average MoE with a pressure of 1.5 MPa is 19586.85, the average MoE with a pressure of 2 MPa is 20055.53 and the MoE is mean with pressure 2.5 MPa is 19391.91. The highest MoE is seen at a pressure of 2 Mpa.

The average tensile strength parallel to the untapped bamboo fiber is 330.93 MPa, the average tensile strength parallel to the fiber at a pressure of 1.5 MPa is 343.84 MPa, the average tensile strength parallel to the fiber at a pressure of 2 MPa is 344.75 MPa and the average tensile strength parallel to the fiber at a pressure of 2.5 MPa are 328.21 MPa. From the results of tensile testing the highest average tensile strength was obtained at a pressure of 2 MPa. From the results of the correlation test, it was found that the tensile strength of the bamboo base on the pressure of the 1.5 felt was significantly correlated with the tensile strength of the bamboo material which was pressed at a pressure of 2.5 MPa. This is because at the pressure of the 2.5 MPa press the bamboo has begun to break down. At press pressure up to 2 MPa does not show a significant relationship because bamboo fiber has not been damaged.

The average shear strength without pressing is 7.55 MPa, the average shear strength at a pressure of 1.5 MPa is 7.66 MPa, the average shear strength at a pressure of 2 MPa is 7.76 MPa and the average shear strength at press pressure of 2.5 MPa is 6.71 MPa. From the test results, it was seen that the highest average shear strength at press pressure of 2 MPa. From the results of testing the partial correlation of shear strength with various variations of press pressure not significantly related. So that the pressure of the press does not affect the magnitude of the sliding bamboo blade. This is because the pressure of the press up to 2.5 MPa does not affect the shear strength parallel to the fiber.

The average compressive strength parallel to the bamboo fiber without pressing is 23.52 MPa, the average compressive strength parallel to the fiber at a pressure of 1.5 MPa is 25.04 MPa, the average compressive strength parallel to the fiber at a pressure of 2 MPa is 27.38 MPa and the average compressive strength parallel to the fiber with a pressure of 2.5 MPa are 27.82 MPa. From the results of these tests it was found that the more affected the pressure of the press higher the greater the compressive strength. From the results of the correlation test it was found that the compressive strength parallel to the fiber at a pressure of 2 MPa was significantly associated with compressive strength parallel to the fiber in the 2.5 MPa pressurization pressure. This is due to the fact that at the pressure of the 2.5 MPa bamboo fiber begins to experience compaction. In testing the perpendicular fiber strength, the highest compressive strength of the bamboo base material was subjected to a pressure of 2.5 Mpa.

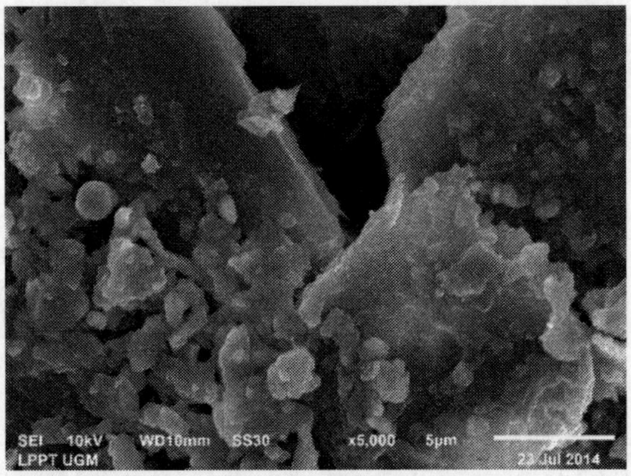

Fig. 4. SEM condition of bamboo fiber at a pressure of 2.5 MPa.

7. Conclusion

Analysis of variants (Anova) of pressure treatment on bamboo material has an effect on some of the mechanical properties of bamboo. Pressure treatments up to 2.5 MPa have a significant effect on the elastic modulus, tensile strength and shear strength of bamboo while the compressive strength is not significantly affected by the pressure treatment of bamboo material. This shows that the strength of the bamboo structure can be used as a substitute for wood construction material.

Bamboo forests have a big influence in reducing the impact of global warming and climate change. Bamboo can be used as an environmentally friendly material and reduce the use of wood as a construction material. Bamboo plants have the ability to cope with natural disasters and their significant ability to absorb carbon in the air gives the future a comfortable and healthy life.

8. References

[1] Gervet, B., 2007, *Deforestation Contributes to Global Warming*, Renewable Energy Research Group, Division of Architecture and Infrastructure, Lulea University of Technology, SE-97187 Lulea, Sweden.

[2] Venkataraman, M., Smitha, 2011, *Causes and Effects of Global Warming*, Indian Journal of Science and Technology, Vol. 4 issue 3, ISSN: 0974- 6846.

[3] Li, Z., Fu M., Xu, D., 2003,*Bamboo Ecosystem and Carbon Dioxide Sequestration*, Journal of Bamboo Research, 22(4): 1-6. (InChinese)

[4] Paulo, M., 2012, *Deforestation Around the World*, Intech, First published March, Croatia, ISBN 978-953-51-0417-9.

[5] Lessard, G., Chouinard, A., 1980, *Bamboo research in Asia*,In: Proceedingsof a workshop held in Singarpore, IDRC, Ottawa, Canada.

[6] United Nations, 1972,*The Use of Bamboo and Reeds in Building Construction*, New York: United Nations, pp 95.

[7] Anon, 1988,*The New Encyclopedia Britannica (Micropaedia)*, vol. 1. Encyclopedia Britannia Inc., Chicago, pp. 980.

[8] Wong, K.M., 1995,*The Bamboos of Peninsular MalaysiaForest*, Research Institute Malaysia (FRIM) in collaboration with Forest Research Center, Forestry Department, Sabah, Malaysia. Malayan Forest Records, No. 41.

[9] Atanda, 2015, *Environmental impacts of bamboo as a substitute constructional material in Nigeria,* Case Studies in Construction Materials, Case Studies in Construction Materials 3 pp. 33–39, Elsevier, journal homepage: www.elsevier.com/locate/cscm.

[10] Alfonso, D.J., 1987,*Let's plant bamboo*,Agribus, Wkly. 1 (19), pp. 25.

[11] Zhou, B., Fu, M., Xie, J., Yang, X., Li, Z., 2005, *Ecological functions of bamboo forest: Research and Application*, Journal of Forestry Research, 16(2): 143-147.

[12] Raka, I. D. N., Wiswasta, I. G. N. A., Budi asa, I. M., 2011, *PelestarianTanamanBambuSebagaiUapayaRehabilitasiLahan Dan Konservasi Tanah Di Daerah Sekitar Mata Air PadaLahan Marginal Di Bali Timur*,Agrimeta, JurnalPertanianBerbasisKeseimbanganEkosistem, Vol. 1, No. 1.

[13] Ghavami, K., (2008), *Bamboo: Low cost and energy saving construction materials,* Proceeding Bamboo Modern Structures, CRC Press, London.

[14] Anonymous, 1997, Healing Degraded Land, INBAR Magazine, 5(3):40-45.

[15] Wu, B., Xie, H., Tan, Shu., 1992,*Preliminary study on water conservation function of Phyllostachyspubescenscommunity*, Journal ofBamboo Research, 1i(4): i8-25. (In Chinese)

[16] Huang, K., Liang, D., Zeng, Z., 1994, Rhizome Distribution of *Phyllostachysmakinio*, Journal of Fujian Forestry College, 14(3): 191-194.(In Chinese).

[17] Chen, K., 2000, Study on Water Conservation Function of Mixed Forests ofChinese Fir and *Phyllostachyspubescens*, Journal of Fujian ForestryCollege, 20(3): 258-261. (In Chinese).

[18] Zheng, Y., Yang, L., 1995,*Hydrologic Effect of Mixed Stands of Cunninghamialanceolata and PhyllostachyspubescensForests*, Journalof Fujian Forestry College, 13(3): 18-22. (In Chinese)

[19] Zeng, L.,Ren, P., Li, Z., 1998, Observation of Biological Characteristicsof *Phyllostachysniduluria*, Economic Forest Research, 16(4):9-11. (In Chinese).

[20] Lin, H., 2002, *Studies on Water Conservation Function of Phyllostachysmakinoiand FokieniahodginsiiMixed Forests*, Fujian Forestry Sci &Tech, 29(3): 18-22. (In Chinese)

[21] White, D. G, Childers, N. E., 1945,*Bamboo for Erosion Control*, J.American Society of Agronomy, 37: 839-847.

[22] Shibata, S., Iwanaga, Y., Kamimura, K., 2002,*Revegetation of Roadside Manmade Slopes with Karami Fencing and by Burying Bambusa multiplexRaeushel Culms*, In: Lou, Y., (ed) Bamboo in disaster avoidance, Beijing: INBAR, 3-11.

[23] Fu Maoyi., Xiao Jianghua., Lou Yiping., 2000,*Cultivation and Utilization on Bamboo*, Beijing: China Forestry Publishing House.

[24] Li, Y., 1998, *Current Status and Prospects of Bamboo Resources in China*, Workshop of developing bamboo and rattan, Beijing, INBAR.

[25] Dong, W., Huang, B., Xie, Z., 2002,*Study on The Pattern of Rhizome Growth of Qiongzhueatumidinodaclone*, Journal of BambooResearch, 21(4): 56-60. (In Chinese).

[26] Xiao, J., 1983, *Subterranean Structure of Timber Bamboo Forest*, Bamboo Research, 2(1): 114-119. (In Chinese).

[27] Wu, B., 1984, *Subterranean Structure and Production Analysis of Moso Bamboo Forest*, Journal of Bamboo Research, 3(1): 49-58. (In Chinese).

[28] Zhou, J., Hu, C., Yang,L., 1985,*Investigation of Rhizomes of Shoot-Oriented High-Yield Bamboo Forest*, Journal of Bamboo Research, 4(1): 57455. (In Chinese).

[29] Wang, K., He, Q., Weng, P., 2000,*Investigation and Analysis on Underground Rhizome and Root System of Phyllostacyspubescensfor Shoot and Timber*, Journal of Bamboo Research, 19(1): 38-43. (In Chinese).

[30] Chen, D., 2002,*Study on The Bamboo Rhizome Growth of PseudosasaamabilisMcCl*, Journal of Bamboo Research, 21(2): 48-53. (In Chinese).

[31] He, J., Fang, W., Lu, X., 1995,*Subterranean Structure of High-Yield and High Effective PhyllostachyspraeconForest for Shoot and Timber*, Journal of Zhejiang Forestry College, 12(3): 247-252. (In Chinese).

[32] Hu, C., Jin, A., Zheng, J., 1994,*Systematic Structure of PhillostachyspraeconRhizome*, Journal of Zhejiang Forestry College, 11(3):264-268. (In Chinese).

[33] Lin, Y., Li, H., Lin P.,2000,*Biomass Structure and Energy Distribution of DendrocalamuslatiflorusPopulation*, Journal of BambooResearch, 19(4): 36-41. (In Chinese).

[34] Lin, Y., Lin, P., Ye, Y., 1998,*Biomass and Structure of Dendrocalamopsisoldhami Population*, Journal of Bamboo Research, 17(2):9-13. (In Chinese).

[35] Xiao, J., 2002,*Pay More Attention to Ecological Benefits of Bamboo Forests*, In: Lou, Y. (ed.), Bamboo in disaster avoidance, Beijing: INBAR, 49450.

[36] Oldeman, L., R., 1990,*World Map of the Status of Human-Induced Soil Degradation*, Wageningen, Netherlands: International Soil Reference and InformationCenter.

[37] UNEP, 1992,*The World Environment 1972-1992 - Two Decades of Challenge*, London: Chapman and Hall.

[38] Hermanto, N. I. S., Eratodi, I. G. L. B., Masdar, A., Morisco, 2010, *PemanfaatanBambuSebagai Material PilihanPadaStrukturBambu Modern*, ProsidingKonferensi Nasional Teknik Sipil 4 (KoNTekS 4), Sanur-Bali, 2-3 Juni 2010.

[39] Yasin, I., Priyosulistyo, H., Siswosukarto, S., Saputra, A., 2015, *The Effect of Variable Lateral Stress of Laminated Bamboo with Artificially Dent Surface on Mechanical Properties*, International Journal of Civil and Structural Engineering– IJCSE, Volume 2 : Issue 2, ISSN : 2372-3971.

[40] Morisco, 2006,*Teknologi Bambu.*Departement of Civil Engingeering and Enviromental, Gadjah Mada University, Yogyakarta.

[41] Ghavami, K., 2005, *Bamboo as Reinforcement in Structural Concrete Elements*, Cement & Concrete Composites, 27 pages 637–649.

[42] Li, S.H., Zeng, Q.Y., Xiao, Y.L., Fu, S.Y. and Zhou, B.L., 1995, *Biomimicry of Bamboo Bast Fiber with Engineering Composite Materials*, Materials Science and Engineering, C3 pages125-130

[43] Xiaohong, G. and Yulong, D., 2005, *Bamboo Science and Culture*, The Journal of the American Bamboo Society,19(1) pp. 16-22.

IOP Conf. Series: Materials Science and Engineering **535** (2019) 012002 doi:10.1088/1757-899X/535/1/012002

Problem analysis on the work cycle of occupational safety and health management system in manufacturing industry

Dianna Ratnawati[1] and S Setuju[2]

[1,2]Mechanical Engineering Education, Faculty of Teacer Training and Education, Universitas Sarjanawiyata Tamansiswa, UH III/1043 Batikan Street, Umbulharjo, Yogyakarta, Indonesia.

E-mail: dianna.ratnawati@ustjogja.ac.id

Abstract. The occurrence of work accidents affecting losses for the industry and workers. Prevention of negative impacts can be controlled through the occupational safety and health management system quality and integrated. This research aims to find the root of the problem in the work cycle occupational safety and health management system using problem tree analysis. This management cycle refers to the OHSAS 18001:2007 standard. This type of research is descriptive qualitative. Based on the results of data analysis, it was found the problems in the occupational safety and health management system at PT. Mega Andalas Kalasan includes the cycle of occupational safety and health policy, planning, implementation and operation, inspection and action to improve occupational safety and health, review of the occupational safety and health management system, and continuous improvement. Problems in each cycle will bring sustainable problems in the next cycle. Findings of problems in the cycle of occupational safety and health policy have an impact on the problems in the initial planning cycle, the weakness of the organization's managerial planning will bring up problems in the implementation and operation cycles, budget constraints in the implementation and operation cycle raises the problem of the lack of real action cycle improvement of safety and health improvement actions work, the absence of formal socialization of audit results to workers in the review and management system of occupational health and safety has an impact on creating barriers to the cycle of continuous improvement.

1. Introduction

Figures workplace accidents in Indonesia are among the highest in ASEAN. Data PT. Jamsostek show until the end of 2010 recorded 0,7% had an accident on the 8,9 million registered workers. Furthermore in 2011 the number of workplace accidents is 48.511 cases in which the biggest cause is the engine, gear transport aircraft and handwork. Impact seen from the lack of control control based on a survey of workplace accidents Executive Director of the Department of Statistics BI Hendy Sulistyowati namely contraction Manufacturing Growth Index (Prompt Manufactuing Index / PMI) in quarter one of 2016 amounted to 46.69% from the fourth quarter of 2015 amounted to 48.23% , This decrease is due to two main factors such as contraction in the total volume of orders and the number of workers.

Zhou et al [1] mentions five factors causing accidents include: (1) management of occupational safety and health; (2) the source of the basic causes; (3) symptoms; (4) contact and (5) loss. More [2] the causes of accidents in the industry, namely the demographic and organizational factors, conditions and work equipment, conditions and unsafe acts, time and place, the health safety and environment training and control measures. Where these factors are part of the cycle occupational safety and health management system. This is in line with the findings of the study [3] that the rate of fatal accidents in the construction industry reduced significantly by implementing occupational safety and health management system. However, some primary and secondary data in the field is just still much evidence of contrast, occupational accidents still occur in many workplaces administratively has passed (comply) audit occupational safety and health management system.

The case of the application of occupational safety and health management system in the garage in the study [4] found in the lack of planning occupational safety and health, implementation, policy, culture occupational safety and health, performance evaluation occupational safety and health needs to be done simultaneously and factors reviewing and improving the performance occupational safety and health needs to be controlled. Furthermore, the findings of the issues [5] PT. kerismas witikco bitung there are discrepancies in the implementation of occupational safety and health management systems. There are five indicators of organizational structure, planning, implementation, procedures, and processes. The indicators included in either category is the planning, implementation and process, while two indicators that fall into the category of less good is the organizational structure and procedures. Likewise, [6] the results of research at PT. Mega Andalan Kalasan that the implementation of the occupational safety and health was not carried out properly, especially from the caretaker committee of occupational safety and Health. The medical examination is only done once a year for workers who work with a high degree of risk in the production unit. Further based on the results of preliminary observations dated August 11, 2017 data showed workplace accidents each year in the PT. Mega Andalan Kalasan experiencing decreased the look of 2015 amounted to 65, 2016 amounted to 52, and in 2017 amounted to 16. However, if examined from the data of occupational accidents, accidents still happen repetition of the same kind each year. By keeping the number of accidents and the repetition of such accidents indicate there are problems in the system of occupational safety and health management at PT. Mega Andalan Kalasan.

In substance occupational health and safety assesment series 18001: 2007 is explained that the work cycle occupational safety and health management system includes the establishment of policies occupational safety and health, planning, implementation and operation, checking and corrective action occupational safety and health, a review of occupational safety and health management system and repairs continuous and integrated in order to create a workplace that is safe, effective and productive [7]. In order to find the problems in each cycle, need to be thoroughly investigated the cause caused the problem. Problem tree analysis, or so-called method is an approach that tree issues detailing a problem into components of the main causes of into order to create a work plan [8]. More [9] issues trees are used to identify and analyze problems by mapping the major cause to its roots and examined result / impact, which means showing the actual state or unexpected situations. On the other hand [10] the problem tree analysis can also identify and visualize a problem and the relationship between the causes of the problem and its effects. It is thus important to identify the problems in the work cycle Occupational safety and health management system in PT. Mega Andalan Kalasan get to the root of the problem and are offset by mapping the impact through the problem tree analysis method to obtain accurate data and analysis.

2. Method

This study used a qualitative descriptive approach in order to gain in-depth and comprehensive study on the issue of the occupational safety and health management system in the industry. The research was conducted at PT. Mega Andalan Kalasan, one of the manufacturing industry in Yogyakarta, Indonesia. The type of data in this study is qualitative data obtained from informants selected by purposive sampling and snowball sampling. Among experts occupational safety and health, chairman of the committee of occupational safety and health, secretary environment health safety, supervisors, and union. Data collection techniques are: (1) in-depth interviews; (2) the participant observation; (3) study the documentation; (4) and a questionnaire. Data analysis techniques using the model of Miles and Huberman [11] including data collection, data reduction, data presentation, conclusion and verification.

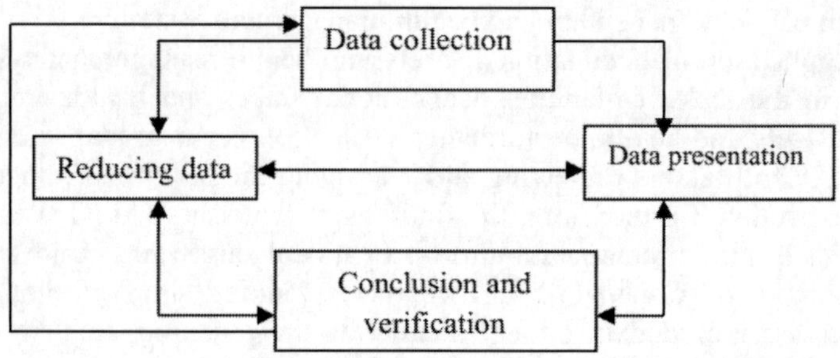

Figure 1. Components in Data Analysis

3. Result and Discussion

3.1. Interview Results

Interviews with the subject of a more detailed study can be seen in Table 1 below.

Table 1. Results of interviews.

Number	Interview Date	Resource	Findings
1.	July 25, 2017	Chairman of the Committee of Occupational Safety and Health	(1) Standard reference in the management of the occupational safety and health management system in PT.Mega Andalan Kalasan using the occupational health and safety assesment series 18001: 2007 (2) Standards of quality assurance based on ISO 9001: 2008 (3) Eenvironmental management system refers ISO 14001: 2004 (4) The structure of organization occupational safety and health management system in concrete simply committee of occupational Safety and Health where the structure is not visible expert First Aid committee and the analysis of the risk / hazard that can handle and prevent occurrence of accident on work (5) The internal audit carried out every year is twice by the management committee of occupational safety and health
2.	July 25, 2017	EHS Officer	(1) There is still work accidents each year despite the decreased number of accidents, as seen from the year 2015 amounted to 65, 2016 amounted to 52, and 2017 totaling 16 (2) Officers there has been no analysis of causes of accidents (3) There are no workplace hazard analysis for each unit of workplace (4) Industrial Parties provide health insurance to workers and are responsible to pay for treatment until cured but there is no First Aid officer in each unit. If there is a work accident immediately rushed to the nearest hospital by

			supervisors of each unit
3.	August 28, 2017	Workers in welding and painting units	(1) The firmness of the industry is lacking in implementation of regulations occupational safety and health while low level of awareness of workers in the use of personal protective equipment , provision of personal protective equipment such as gloves and masks less, the handling of the incident of workplace accidents less nimble by the head of personnel, frequent delays in the supply of the contents of the first aid box, the industry is responsible for financing the treatment of employees who are injured at work, there is always a training safety and health for employees for each year (2) There are no specific reports of performance management systems occupational safety and health to be studied as a material management system improvement safety and health but routine audit of internal audit has been carried out.

3.2. Observations

Based on observations carried out at the production unit of PT. Mega Andalan Kalasan on August 3, 2017 obtained a description that working conditions at PT. Mega Andalan Kalasan has been coordinated with neat and well as: (1) the technician is always standby and work according to the job; (2) management and placement of goods as well as regular equipment equipped with a sign clan; (3) lighting and temperature well equipped facilities and fan lights corresponding quantity; (4) the level of cleanliness and tidiness of the environment very well seen from the actions of workers always disciplined cleaning and tidying work area after the work is completed and provided a waste disposal plant; (5) regular medical examinations on a regular basis each year as evidenced by the reports of medical examinations for workers; (6) solid teamwork between labor and supervisors to provide oversight and guidance; (7) occupational safety and health facility complete with their visible signs occupational safety and health, fire extinguishers (light fire extinguishers, hydrants, fire alarms), sign hazard.

However there are some findings that deviate include: (1) the noise level is not good, especially in the cartor unit but not matched with the use of ear muff; (2) not availability of a suggestion box for workers; (3) personal protective equipment provided good helmet, ear muff, gloves, masks etc but there is still negligent workers do not wear gloves, helmets, goggles and boots while performing the work; (4) boxes occupational safety and health is provided but it looks less complete only gauze, betadine, bandage and alcohol.

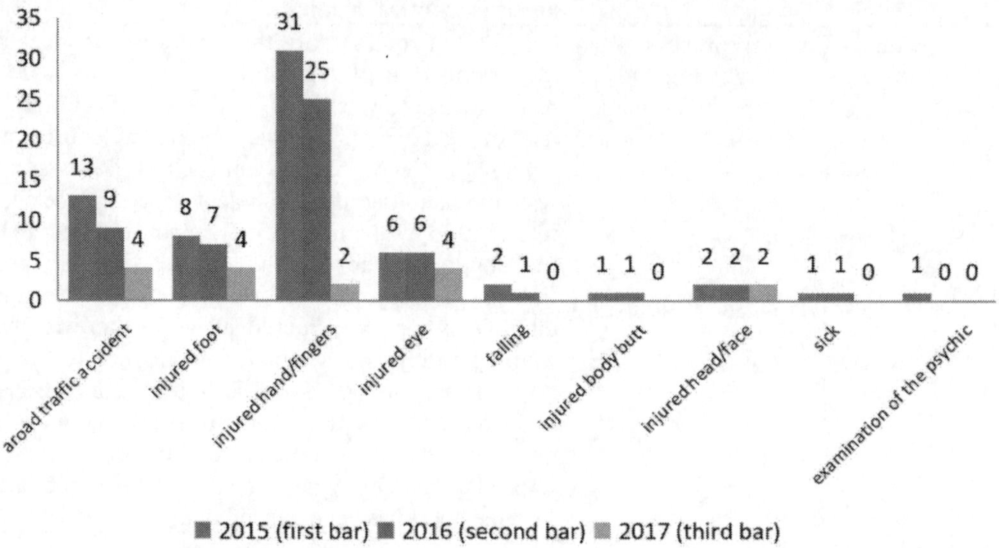

Figure 2. Graph accident work frequency 2015-2017.

Figure 3. Activity and good working environment.

Figure 4. Activity of labor and facility completeness occupational safety and health unfavorable.

3.3. Questionnaire Results

Data from the questionnaires has been reduced by the method of problem tree analysis, can be seen further in Table 2 below.

IOP Conf. Series: Materials Science and Engineering **535** (2019) 012002 doi:10.1088/1757-899X/535/1/012002

Table 2. Results of the identification problem Occupational safety and health management system with problem tree analysis method

Number	Identify the problem	Findings
1.	Occupational Safety and Health Policy	Cause : (1) Commitment of management and performance occupational safety and health less consistent, as indicated by the lack of attitude / strict sanctions against negligent workers wear protective equipment (2) Occupational safety and health policy has been communicated to the workers but awareness of the obligations of occupational safety and health less (3) Less input from workers because the absence of a suggestion box as a conduit criticism and suggestions from workers to reporting the results of the evaluation / field conditions (4) Preliminary review conditions of occupational safety and health less: provision of facilities for personal protective equipment for masks and gloves one week only twice and not provided eyeglasses; Keith their work accident risk control through the analysis of causation and prevention of occupational accidents handled by a competent human resources; provision of drug / first aid box contents is often too late. Result : (1) The proliferation of undisciplined work culture (2) Occurred repetition of the same accident (3) Occurred dual role by supervisors that is to guide and coordinate the work process in accordance unit of work and take care of the cost of accommodation / treatment of workers injured at work
2.	Planning	Cause: (1) Management lacking organization, socialization legislation / regulations occupational safety and health to-date has not been communicated to workers (2)Constrained by the administration of the financial management of the occupational safety and health: claim funds referral treatment of workers to the hospital long descent (4) Planning is the occupational safety and health only involves the committee development occupational safety and health, yet their analysis of accidents / only the work accidents (5) Plan of occupational safety and health is less aligned with the national legislation (6) a risk assessment has not been through procedure hazard identification. Result: (1) the absence of sanctions which according to the rules expressly legislation for employees who violate the rules of occupational safety and health (2) Less full contents of the box occupational safety and health in accordance legislation (3) Figures have not reached zero accident accident
3.	Implementation and Operation	Cause: (1) Budget has not memadahi, fulfilling tough safety equipment / has not been realized as a whole, such as the availability of glasses has been no cause accidents "eye contact with gram" repeated annually, the engine in unit maintenance there are worn not been replaced but only manoeuvered so that production remains the (2) lack of availability of human resources responsible for matters of Safety and working in the field of health, especially that dealing with emergencies such as accidentswork (3) Not yet reporting and documentation of the work habits is not correct, the recommended discipline occupational safety and health has not been enforced in all

		work units, orderly if there is supervision / imaging esp usnya in unit maintenance is warning / command to wear a helmet if there is a guest (4) management infrastructure is less well-run, regular maintenance / repair / replacement work equipment should be observed to control accidents, such as drilling machines in the export oriented product unit have been unsuitable result thumb injury as the drill; The supply chain management unit goods storage rack unsuitable cause head injury. Air pollution levels in HE welding and painting units not managed optimally. (5) Workers' awareness of the importance of working according to standard operating procedures and occupational safety and health as well as concern for the safety co-workers need to be improved, because in unit maintenance work accident occurred because the third party / other workers. (6) Participation of workers less involved for investigation of incidents of accidents and hazards identification and the determination of control accidents. Result: (1) availability of protective equipment has not been evenly distributed to each unit of production (2) occurred accidents that recur (3) Working tools / work facility into causes of accidents (4) occurred workplace accidents because running a work instruction that is less right
4.	Inspection and corrective action Occupational Safety and Health	Cause: (1) Audit occupational safety and health held twice each year, but the results have not been used for corrective actions occupational safety and health in good repair facility shortage occupational safety and health, maintenance tools, as well as the investigation of incidents work accident. (2) Examination of the performance of the occupational safety and health in each work unit was uneven due to lack of human resources, the organization has not proactively monitor compliance with occupational safety and health management system program and operational criteria are further documented in the occupational safety and health management system. Result: (1) Facilities occupational safety and health has not fulfilled the quality and quantity (2) There has been no in-depth study analysis of workplace accidents (3) There has been no improvement of the organizational structure occupational safety and health management system (4) Planning occupational safety and health did not cover all the needs in each work unit (5) Commitment the organization has not consistently related to compliance with the legislation occupational safety and health.
5.	Review of the Occupational Safety and Health Management System	Cause: (1) The results of a review / audit occupational safety and health management system has not been formally communicated to workers (2) Repairs and improved performance occupational safety and health management system is not optimum, changes in the organizational structure but the mapping task is not yet clear, the contribution of the worker role for performance improvement occupational safety and health management system is not visible because we are not included in the forum planning occupational safety and health, accident data only recapitulated of pending work for prevention strategies (3) The management review have not considered input / complaints in each unit. Result: (1) There has been no improvement of the board and employees in relation to the working attitude which deviates from the

		responsibilities and regulations occupational safety and health (2) A review occupational safety and health management system just in writing, no direct action as a preventive for performance improvement occupational safety and health management system in thoroughly.
6.	Continual Improvement	Cause: Improvements are still in the stage of the administrative / written evaluation Result:Accidents still happen with three contributing factors: human error,equipment, and the environment.

Based on the study results of the study, they found the problem / misbehavior of occupational health and safety assesment series 18001: 2007. Such irregularities can be identified: (1) the policy occupational safety and health; (2) planning; (3) the implementation and operation; (4) checking and corrective action occupational safety and health; (5) review of occupational safety and health management system; and (6) continuous improvement. Duty cycle occupational safety and health management system will be mutually affect each other's performance, because it is a coherent and integrated. As well as the problem of lacking commitment to the industry occupational safety and health policy, will have an impact on the planning cycle that is defective managerial and administrative organizational structure of financial management. Managerial awareness amongst the general manager to the manager occupational safety and health to develop a occupational safety and health management system effect on the cycle of continuous improvement activities [3].

On the other hand in the planning cycle issues such as the lack of analysis of hazards in the workplace and there is no hazard control officers through the analysis of causes of accidents will impact the repetition of the same accident as the data in Figure 2. Hazard analysis can begin by determining factor of danger in the workplace ranging from distinguished as dangerous (physical condition), harmful (chemical, biological, and physical) and cumbersome (psychophysical) in which the whole of these factors can potentially disease and accidents [12]. Accidents will impact more harmful to workers than companies. Losses incurred such workers financial, physical and psychological consequences far more numerous and longer [13]. Buoyed by the post-crash inspection data psychic by the PT. Mega Andalan Kalasan. While controlling hazards in the workplace can be done through continuous improvement actions in cycle occupational safety and health management system. Recommendations for corrective actions can be done through the goal-setting phase, an approach, establish procedures and conduct continuous evaluation [14].

Problems on the application and the operating cycle as indicated by the lack of human resources field of occupational safety and health will also affect the cycle of checking and corrective action as well as the response / reporting the number of completeness first aid facilities in the work unit. Key to the effectiveness of the implementation of occupational safety and health management systems depends on the type of system used, the senior management commitment and employee participation are effective. Employee engagement can be enhanced through the attitude of senior manager, management commitment, communication between workers and managers, the initiative / commitment to worker participation, knowledge and training baigi workers, providing suggestion boxes and performance evaluation occupational safety and health management system is more engaging employees [15] [16]. Further to the internal audit team can be based on standards and criteria occupational safety and health management system and conducted by members who are skilled to be actualized in the act of continuous improvement so that success in implementing occupational safety and health management system can be achieved, especially in the control of occupational accidents [17]. While the most influential factor in improving the implementation of occupational safety and health management systems is first aid, rest areas, waste management, toilets, personal protective equipment, labor, and signposts occupational safety and health [18]. Where these factors part of the policy cycle occupational safety and health, planning, implementation and operation.

4. Conclusion

Based on the results and discussion of research, it can be concluded that the problems that occur during each cycle occupational safety and health management system can be mapped to: (1) the policy occupational safety and health needs to scrutiny re-examine their initial conditions occupational safety and health, the policy commitments occupational safety and health and less powerful and less involving input / complaints of workers in policy formulation occupational safety and health; (2) planning is still weak in managerial organization, lack of attention to legislation that is relevant, and the lack of analysis of hazard control; (3) the implementation and operation are constrained in the budget issues occupational safety and health, the management of facilities and infrastructure on each unit of work, lack of awareness of workers on the occupational safety and health and the limited number of human resources in the organizational structure; (4) checking and corrective action issues arise there has been no real action for the improvement of audit results occupational safety and health Management System, the performance inspection occupational safety and health has not been evenly distributed on each unit; (5) review of occupational safety and health management system found the problem has not been lack of socialization audit results occupational safety and health management system formally to the worker; (6) the continued improvement in the phase still administrative / written evaluation.

Acknowledgments
The author successful in completion of this research doesn't slip out of the role of various parties. Authors would like to thank you for research grants that have been given by the direktorat riset dan pengabdian kepada masyarakat, direktorat jendral perguruan tinggi. And also the author goes to PT. Mega Andalan Kalasan who has given permission for the good research data retrieval and sharing of problems and solutions in the field.

5. References

[1] Zhuo S, Danlong L, Minggang L, and Mingrui X 2015 The Establishment and analylis of accident-controlled model based on three-dimensional coordinates. *The open automation and control systems J.* 7 p 1034
[2] Mohammadfam I, Soltanzadeh A, Moghimbeigi A, Akbarzadeh M 2014 Factors affecting occupational accidents in the construction industry (2009-2013). *JOHE, Spring.* 3 p 88
[3] Yoon S, Lin H, Chen G, Yi S, Chio J, Rui Z 2013 Effect of occupational health and safety management system on work-related accident rate and differences of occupational health and safety management system awareness between managers in south korea's construction industry. Saf health work A. 4 p 201
[4] Hidayat N, and Whyuni I 2016 Kajian keselamatan dan kesehatan kerja bengkel di jurusan pendidikan teknik sipil dan perencanaan fakultas teknik UNY. *Jurnal Pendidikan Teknologi dan Kejuruan.*25 p 51
[5] Woun A 2013. Analisis penerapan sistem manajemen keselamatan dan kesehatan kerja PT. Kerismas witikco bitung. *J UNSTAT*

IOP Conf. Series: Materials Science and Engineering **535** (2019) 012002 doi:10.1088/1757-899X/535/1/012002

[6] Martin T 2006. Studi penerapan keselamatan dan kesehatan kerja pada perusahaan alat kesehatan PT.Mega Andalan Kalasan Yogyakarta (Purwokerto: Perpustakaan kampus7)

[7] OHSAS 18001: 2007. Occupational Health and Safety Management Systems-Requirements

[8] Miller P 2004 private communication (Mr. Miller was a consultant at Deloitte Consulting, where they Widely trees ude issue)

[9] Scavarda et al, 2004. A review of the causal mapping practice and research literature. second world conference on POM POM and 15th annual conference (Mexico: cancun)

[10] Ammani et al, 2010. Applying the problem tree analysis methodology to the ADP system in Nigeria. Agricultural extension J. 14

[11] Miles and Huberman M A 1994. Qualitative analysis of data. California: SAGE Publications Inc.

[12] Hola B, Nowobilski T, Szer I, J Szer 2017. Identification of factors affecting the accident rate in the construction industry. *Procedia*engineering. 208 p 35

[13] Hrymak V and Perezgonzalez J 2007. The costs and effects of workplace accidents twenty case studies from ireland. *Health and safety authority research series 02.*

[14] Hargiyarto p 2011, Analisis kondisi dan pengendalian bahaya di bengkel/laboratorium Sekolah Menengah Kejuruan. *Jurnal Pendidikan Teknologi dan Kejuruan*. 20 p 203

[15] Gallagher C, Underhill E, and Rimmer M 2001. Occupational health and safety management systems: a review of Reviews their effectiveness in securing healthy and safe Workplaces (Sydney: National Occupational Health and Safety Commission)

[16] HSE- Health and safety Executive, 2005. Obstacles Preventing worker involvement in health and safety. Prepared by ECOTEC Ltd for the health and safety axecutive

[17] Pillay M 2015. Accident causation, prevention and safety management: a review of the state-of-the-art. *Procedia manufacturing.* 3 p 1838

[18] Rush I, Hamzah B, and Asdar M 2017. Implementation of the management systems on the safety and health to construction workers. *International Journal of Engineering Inventions.*6 p 35

IOP Conf. Series: Materials Science and Engineering **535** (2019) 012003 doi:10.1088/1757-899X/535/1/012003

The implementation of teaching factory in producing fly ash based eco-friendly concrete products

NA Inderawan[1] and S Widodo[2]

[1]Pascasarjana, Yogyakarta State University, Caturtunggal, Depok, Sleman, 55281, Yogyakarta, Indonesia
[2]Study Program of Civil Engineering and Planning Education, Faculty of Engineering, Yogyakarta State University, Caturtunggal, Depok, Sleman, 55281, Yogyakarta, Indonesia

E-mail: noorarifin1309@gmail.com

Abstract. This literature study aimed to describe eco-friendly solutions for developing teaching factory models in vocational high schools by using Fly Ash wastes as cemet replacement materials. The need for eco-friendly concrete has become an innovation which is able to reduce the impacts of environmental pollutions and the concrete product has better quality than conventional concrete. Vocational High Schools in Indonesia have good potential to apply this idea by using teaching factory model for producing eco-friendly concrete because of multiple benefits: (1) reducing the amounts of fly ash wastes; (2) the result of this model is a good quality concrete product that gives benefit to society; (3) providing the understanding of Greening TVET to vocational high students; (4) students deepen their competencies by adapting Greening TVET; (5) also growing good cooperation between steam electricity power plant, vocational school, industry in the field of eco-friendly concrete, and market needs. This study reviewed the various studies about the characteristics and optimum mix design of fly ash as material mixture and teaching factory model. The result of study are idea and design of teaching factory model in Vocational High School by using fly ash as cement replacement of concrete mix design.

1. Introduction

The handling of fly ash waste by Steam Electricity Power Plant is still a worrying environmental problem which continues to increase in line with the increasing demand for coal as a source of electrical energy for the society. The amount of Steam Electricity Power Plant of Suralaya, Banten, produces coal waste approx. 2.7 million tons per year and predicts to grow up to 11.2 million tons/year on 2027 [1] also in the Asam-Asam Electricity Power Plant, a pile of coal ash waste burning two plants reaches 130.000 tons [2]. In Indonesia, the disposal and handling of this waste is still limited to stockpiling on vacant land. The waste of fly ash is hazardous waste, and if not managed properly, will cause environmental pollution problems. Solution of the problem is by using fly ash which obtains high silica content and pozolanic properties as cement replacement materials in producing concrete products. Fly-ash-based concrete proved to be good in structural applications which has similar strength to those of conventional cement concretes. Sarker et. al has found that fly-ash-based geopolymer concrete could reach higher strength than conventional cement concrete [3,4].

The use of fly ash waste as partial replacement of concrete products is an application of Greening TVET for covering Green Jobs which regulates limiting greenhouse gas emissions and reducing the amount of emissions of contaminated waste [5, 6]. This concept is for Vocational education to be able to run vocational education in preserving the environment and reducing emissions of contaminated waste. If fly ash is used as a manufacturing material and substitute material on concrete, it will reduce the amount of Portland cement usage. The continuous use of Portland cement in producing concretes can contribute CO_2 emissions in increasing the risk of global warming.

Implementation of that idea can be applied to Teaching Factory model in Vocational schools. Teaching factory model or Learning factory was known as development of production units related to industry in which the process of production or service activities refers to the curriculum in vocational schools [7]. This model integrates a practice-based curriculum and the provision of facilities to create products [8] where students are given open access to the industry in active learning laboratories where students are provided with an effective cost understanding and new experiences about the reality of design, manufacturing and business practices [9]. Teaching factory is expected to support knowledge-based manufacturing by develop technologies and skills to improve product/process innovation [10]. The innovation could be eco-friendly concrete products by using alternative materials such as an unused material, which if used, could reduces the amount of contaminated waste and Portland cement.

Researches focused on alternative materials based as cement replacement has been studied. Proske et.al [11] experimented by using limestone powder as cement replacement materials and found that it could reach compressive strength about. 40 N/mm^2. Termkhajornkit et. al. found that the increase of compressive strength after 28 days of concrete pastes containing 25% fly ash was higher than those of conventional cement paste, but at 50% ratio of fly ash, the strength was decreased [12]. In the same way, Sathawane et. al. experimented by using 22.5% fly ash and 7.5% rice husk ash in several different mixed proportions by replacement of cement in concrete, could get maximum compressive strength, even at more than those ratio are used, the compressive strength decreased [13]. Based on those studies, eco-friendly concrete could be an alternative solution as a normal concrete, but the use of alternative material is only served to reduce the use of cement, still limited to a certain degree where the mix design ratio of materials used can achieve the optimum compressive strength and workability of the concrete. However, the implementation of this model is not easy and requiring good coorporation between Steam Electricity Power Plant, Vocational High schools, Industry partners and market needs. The use of fly ash waste as cement replacement materials in the application of teaching factory model will reduce the amount of environmental emissions and could deepen students' competence in the application of teaching factory model and understanding of Greening TVET idea.

2. Characteristics of Fly Ash

Fly ash is waste carried out during combustion and captured by electrostatic precipitator which consists 58.75% SiO_2, 25.82% Al_2O_3, 5.30% Fe_2O_3, 4.66% CaO, 1.36% alcali, 3.30% MgO and 0.81% other materials [14]. Even it's categorized as hazardous waste, the Class F fly ash is potentially good to use because of pozzolanic properties as cement partial replacement [15] which could improve the long term durability of eco-friendly concrete by reducing entrance of chloride ions [16]. Also, partial replacement of cement with fly ash could reduce production costs [17] because it reduces the cost of purchasing cement and the supply from Steam Electricity Power Plant just needs the usage permission letter, eventhough at the end process, Vocational High School should give the report about the detail of fly ash wastes usage. Fly ash can be used as filler cavity and fastener between aggregates because of very soft particle size, also has strong adhesion advantages because it contains silica and alumina with low lime content [18]. Arifin et. al [19] found that the fly ash filler could get the high-stability pavement and the more rigid will have a low deformation capability.

Fly ash has good characteristics as a cement replacement material, but only in certain degree. Koukouzas et. al [20] found that the use of fly ash about 20% as cement replacement, was able to reach compressive strength up to 25 MPa and flexural strength up to 3.4 MPa, higher than those of non-fly ash standard blocks. Likewise Termkhajornkit et. al. [12] founded that 25% fly ash as optimum ratio for reaching the maximum compressive strength. Dhiyaneshwaran et. al [21] indicates the use of fly ash as partial replacement could increase of compressive strength of concrete starting from the use of 0%, 10%, 20% and 30% fly ash instead of cement, but after the fly ash 40%, there is a decrease of compressive strength. In addition, Darmawan [22] found that with a mixture of coal ash which is a Steam Electricity Power Plant wastes, was successfully made brick with a simple way and concluded that bricks by using coal ashes with 60:40 ratio (68.98 kg/cm^2) higher than ordinary brick (50.45 kg/cm^2). It means that the use of fly ash has an optimum limit for concrete to achieve good workability and compressive strength. The use of fly ash does not completely replace cement as the main binder, but is used only in part where certain degree of fly ash reach workability and the optimum compressive strength is good for concrete.

3. The concept of teaching factory and mix design in producing eco-friendly concrete products

Implementation of teaching factory in vocational schools could deepen students' understanding in the field of practice-based concentration on actual problem in industry and working based on industry standard [23]. Chryssolouris et. al [24] stated that teaching factory model was scheme to transfer knowledge from vocational school to industry which industrial grade installed to vocational school facilities, so it could be used as demonstrator for new technological concepts to be validated by students. The implementation of teaching factory model in vocational schools in Indonesia is important to strengthen cooperation between vocational schools and industries. Vocational High School is obliged to cooperate with Business and Industry with the aim that the education program held and its graduates in accordance with the needs of labour market [25]. Every vocational high schools must have a binding relationship with industry or labour market for all learning stages from planning, implementation, learning process, supervision, assessment and certification [26].

The use of fly ash wastes requires the support and cooperation of the various parties involved. Fly ash is categorized as hazardous waste in Indonesia, which it may be used by vocational schools if the Government of Indonesia permits for the use and distribution to a predetermined rate. If the usage permission of fly ash has been approved, then the steam electricity power plant, vocational high schools, and related industries can work together in production eco-friendly concrete. Process is started from Vocational High School which proposed the amount of fly ash waste needs in detail of production material, so Steam Electricity Power Plant could supply them. Steam Electricity Power Plant could be involved in monitoring the process of eco-friendly concrete production in Vocational High School. Vocational school as production site applying teaching factory model must provide mix product design, applying occupational health and safety to its production process, in cooperation with industry engaged in eco-friendly concrete. Cooperation between vocational high school and industry in the form of supplying tools and machines of good quality by industry, production process and management of occupational health and safety by using quality standards applied by the industry, and the final process of distributing eco-friendly concrete products to society. The society as the customers are obliged to give feedback in the form of opinions, complaints or suggestions to Vocational High School, Industry and Government related to production of eco-friendly concrete about the quality and the satisfaction when using the products. This feedback is reviewed and then become suggestions of improvements towards the eco-friendly concrete production process in vocational high school. The detail method of Teaching Factory model concept used in Vocational High School adapted from Teaching Factory Model Guided Book [27] as follows:

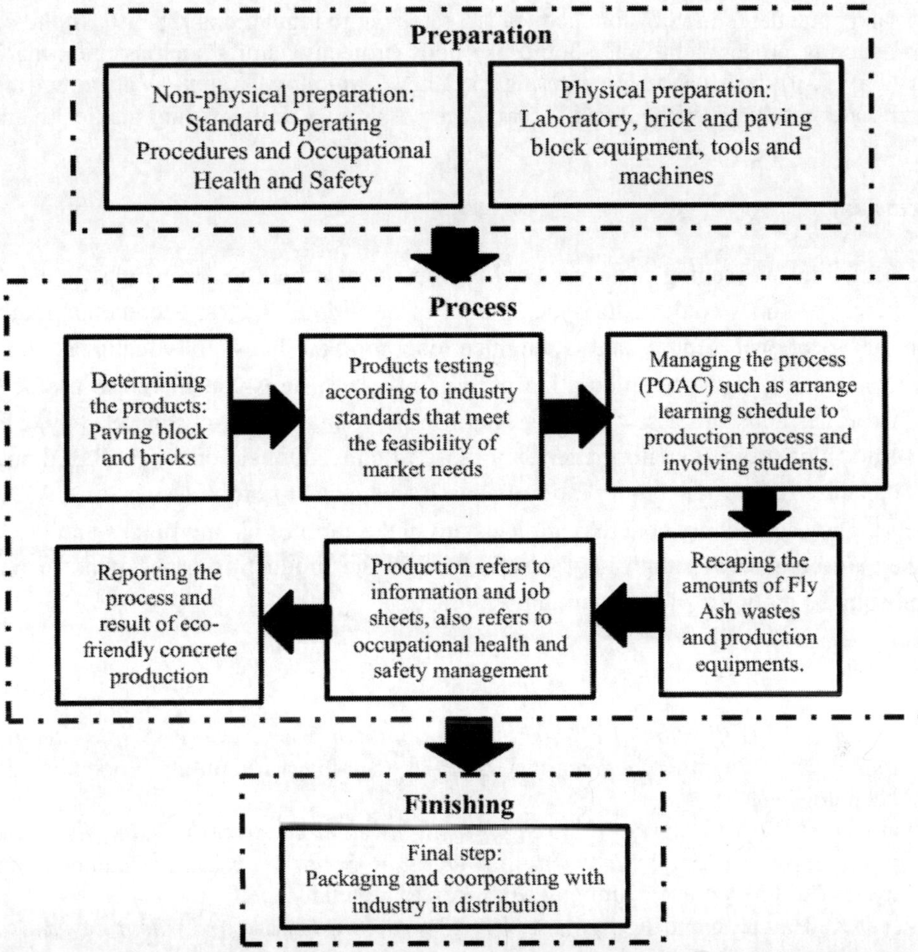

Figure 1. Process in producing fly ash based eco-friendly concrete in vocational high school

The eco-friendly concrete production process begins by determining the type of product and mix design which is appropriate for the society's needs. The products do not have to be luxurious but simple and could be used by all society and the manufacturing process is not complicated and wasteful. Products could be paving blocks or bricks. The ratio of the use of fly ash content in concrete mix design as cement replacement material is adapted from mix design ratio of Dhiyaneshrawan et. al [21] to reach maximum compressive strength. Mix design of eco-friendly concrete products is shown in table 1. as follows:

Table 1. Mix design concept of eco-friendly concrete products (paving block and bricks)

No.	Eco-friendly products	Concrete mix-design/compositions ratio.
1.	Concrete paving blocks	1 Binder (70% Portland cement + 30% Fly ash) : 3 Fine Aggregates : 5 Coarse Aggregates

2.	Concrete bricks	1 Binder (70% Portland cement + 30% Fly ash) : 4 Fine Aggregates

Implementation of production should be well supervised and paying attention to occupational health and safety standards in vocational education referring to regulations [28, 29]. In the eco-friendly concrete production process, students should use body protective stuffs such as thick masks and eye protection to avoid fly ash waste for entering students' respiratory system. Vocational high schools and industries are absolutely responsible for providing assistance and ensuring the health and safety of students.

4. Conclusion

The implementation of teaching factory model in vocational school in producing fly ash based eco-friendly concrete products could reduce the amount of hazardous wastes, become alternative product innovation of vocational school and strengthen coorporation between vocational school, steam electricity power plant, industry, and market needs. Fly ash waste is considered to use because of its pozolanic properties and with a certain usage ratio in the manufacture of concrete, fly ash can reach the maximum compressive strength, better than those of conventional concretes. Based on the results of various literature review, the binder ratio of fly ash and cement in mix design is used (30% fly ash: 70% cement). The eco-friendly concrete products are in the form of paving blocks and brick products, because they are considered simple, not complicated in the production process, not expensive, and easily applied in the daily life of the community.

5. References

[1] Thahir Z A 2017 *Utilization of Fly Ash and Bottom Ash from Suralaya Steam Electricity Power Plant Banten to manufacture of GEOPAV* (Surabaya: Sepuluh Nopember Institute of Technology)

[2] Haryanti N H 2015 *Compressive strength of lightweight concrete by using fly ash admixtures from Asam-Asam Steam Electricity Power Plant, South Kalimantan* (Banjarmasin: Paper of Physics Study Program, Lambung Mangkurat University) p 20

[3] Sarker P K, Haque R and Ramgolam V 2012 *Fracture behaviour of heat cured fly ash based geopolymer concrete* Materials and Design vol 44 p 585

[4] Sarker P K 2011 *Bond strength of reinforcing steel embedded in geopolymer concrete* Materials and Structures vol 44 p 1027-1029

[5] UNESCO-UNEVOC 2013 *Greening TVET for Sustainable Development* p 14

[6] UNEP, ILO, IOE and ITUC 2008 *Green Jobs: Towards decent work in a sustainable, low-carbon world* (Washington, DC.: Worldwatch Institute)

[7] Sutopo A 2013 *Potential For Production Units To Improve The Working Competence Of Vocational School Students* Journal of Technology and Vocational Education vol 14 no 1 (Medan: Medan State University) p 77-87

[8] Lamancusa J S, Jorgensen J E, Zayas-Castro J L, De Ramirez L M 2001 *The Learning Factory-Integrating Design, Manufacturing and Business Realities into Engineering Curricula - A Sixth Year Report Card* International Conference on Engineering Education p 1

[9] Lamancusa J S and Simpson T W 2004 *The Learning Factory - 10 Years of Impact at Penn State* International Conference on Engineering Education Gainesville, Florida p 1

[10] Chryssolouris G, Mavrikios D and Mourtzis D 2013 *Manufacturing Systems: Skills and Competencies for the Future* Procedia CIRP vol 7 p 23

[11] Proske T, Hainer S, Rezvani M and Graubner C A 2013 *Eco-friendly concretes with reduced water and cement contents – Mix design principles and laboratory tests* Cement and Concrete Research vol 51 p 38-40
[12] Termkhajornkit P, Nawa T, Yamashiro Y and Saito T 2009 *Self-healing ability of fly ash-cement systems* Cement and Concrete Composites vol 31 no 3 p 195-203
[13] Sathawane S H, Vairagade V S and Kene K S 2013 *Combine Effect of Rice Husk Ash and Fly Ash on Concrete by 30% Cement Replacement* Procedia Engineering vol 51 p 35-44
[14] Munir M 2008 *The use of fly ash waste in producing high quality and safety hollow block for environtment* (Semarang: Graduate School in Environmental Science of Diponegoro University)
[15] ASTM Standard C618-08a 2008 *Standard specification for coal fly ash and raw or calcined natural pozzolan for use in concrete* (West Conshohocken, PA: ASTM International)
[16] Nath P and Sarker P 2011 *Effect of fly ash on the durability properties of high strength concrete* Procedia Engineering vol 14 p 1155
[17] Dilmore R M and Neufeld R D 2001 *Autoclaved aerated concrete produced with low-NOx burner/selective catalytic reduction fly ash* Journal of Energy Engineering vol 127 p 37
[18] Wardani and Retno S P 2008 *Utilization of coal waste (Fly Ash) for soil stabilization as well as other civil engineering needs in reducing environmental pollution* (Semarang: Faculty of Engineering Diponegoro University)
[19] Arifin M Z, Sadillah M and Wicaksono A 2018 *The influence of the use of fly ash as filler toward Modulus recience of Asphalt Concrete-Wearing Course* Journal of HPJI vol 4 p 59-66
[20] Koukouzas N, Papayianni I, Tsikardani E, Papanikolaou D and Ketikidis C 2007 *Greek fly ash as cement replacement in the production of paving blocks* (Northern Kenthucky, USA: Journal of World of Coal Ash) p 7-10
[21] Dhiyaneshrawan S, Ramanathan P, Baskar I and Venkatasubramani R 2013 *Study on durability characteristics of Self-Compacting Concrete with fly ash* Jordan Journal of Civil Engineering vol 7 no 3 p 348
[22] Darmawan A D 2008 *The ash of absorbent waste* (Jakarta: Journal of Science and Technology)
[23] Rentzos L, Doukas M, Mavrikios D, Mourtzis D and Chryssolouris G 2014 *Integrating Manufacturing Education with Industrial Practice using Teaching Factory Paradigm: A Construction Equipment Application* Procedia CIRP vol 17 (Greece: Laboratory for Manufacturing Systems and Automation, University of Patras) p 189-194
[24] Chryssolouris G, Mavrikios D and Rentzos L 2016 *The teaching factory: A Manufacturing Education Paradigm* Procedia CIRP p 44-48
[25] Regulation of the Minister of Education and Culture of Indonesia no 36 2014 on *Guidelines for the Establishment, Change and Closure of Schools*
[26] Government Regulation of Indonesia no 41 2015 *Industrial Resource Development*
[27] Sub Directorate of Curriculum, Directorate of Vocational Education Directorate General of Primary and Secondary Education, Ministry of Education and Culture of Republic Indonesia 2017 *Teaching factory implementation guide* p 34-39
[28] Regulation of the Minister of Manpower and Transmigration no 01 1980 *Occupational Health and Safety on Building Construction*
[29] Joint Decree of the Minister of Manpower and Minister of Public Works 1986 *Occupational Health and Safety in Construction Places*

IOP Conf. Series: Materials Science and Engineering **535** (2019) 012004 doi:10.1088/1757-899X/535/1/012004

The concept of computational thinking toward information and communication technology learning

R Harimurti[1], Ekohariadi[2], Munoto[3] and IGP Asto B[4]

[1]Informatics Department, Surabaya State University, Surabaya, Indonesia
[2,3,4]Electrical Engineering Department, Surabaya State University, Surabaya, Indonesia

E-mail: rinaharimurti@unesa.ac.id

Abstract. Computational thinking is a process of problems solving and designing systems using concepts in computer science. Computational thinking is based on a computational principle that involves a wide range of approaches and skills with applications in several disciplines. Computational thinking has been widely applied in the curriculum of various levels of education while still basing on the principles and characteristics of a strong computational thinking. To train the computational thinking to students it must include the computational thinking material in the compulsory material. This material needs to be held since the level of basic education with a simple problem level, the level of the material will increase along with the increase of education level. This article discusses the implementation of computational thinking on information and communication technology learning using a simple programming language that is Scratch. This programming language will make it easier for students to understand the concept of programming by using blocks so that students will easily learn. As a basis for information and communication technology learning, a measurement scale has been developed to measure students' computational thinking ability. Computational thinking uses a five-point Likert scale consisting of 29 items from 5 developed indicators.

1. Introduction

Learning is an activity that involves a person in an effort to gain knowledge, skills, and positive values by utilizing various sources for learning. Learning can involve two parties, namely students as learners and teachers as facilitators or teachers. The most important thing in the learning activities is the learning process (learning process).

Current technological developments have made computer devices a very effective learning tool, as they are widely used for routine tasks, helping to solve complex problems as well as designing systems. The basic principle of problem-solving techniques and their application is the concept of development of computational thinking (CT).

Computational Thinking (CT) is essentially problem-solving, designing systems, and understanding human behaviour, drawing on fundamental concepts to computer science can be performed by humans or machines (computers) [1] CT has an important role in computer application development, but CT can also be used to support problem solving in all disciplines, including humanities, mathematics and science. Currently at all grade levels, CT competence has been applied as it is a basic principle of problem-solving techniques and their application. Programming is more than coding, as it brings students to computational thinking that involves solving problems using computer science concepts such as abstraction and decomposition [2] Even for non-computing majors,

CT is applicable and useful in everyday life. Not surprisingly, having such capability is a must for someone living in the digital age today.

This article will discuss how the conceptual framework of CT underlies the development of ICT learning. Section 2 discusses the methodology, and Section 3 is the results and discussion. Finally, Section 4 summarizes the research.

2. Method

Literature study is used in the compilation of this article, this method is done by collecting references, books and journals related to the discussion of theory, analysis, research that has been done and also the implementation of CT on ICT learning. The journal used as a source begins from 2006 to the latest in 2017. The journals used for reference can be grouped as follows: basic concept of CT, CT development, curriculum developed based on CT and CT implementation on various levels of education. Keywords in the search of journals that support the writing of this article is Computational Thinking (CT).

3. Result and Discussion

3.1 Basic Principles of Computational Thinking (CT)

Computational thinking (CT) is based on a computational principle that is a term that covers a variety of conceptual and engineering tasks, so computational thinking involves a broad set of approaches and skills with applications in multiple disciplines. Computational thinking (CT) developed by Jeannette Wing [1] to describe a set of thinking skills, habits and approaches that are integrated to solve complex problems using computers. CT developed several key principles [3] which are listed below. (1) how to solve problems and design the system refers to the basic concepts on computer science (2) create and use different levels of abstraction to understand and solve problems more effectively (3) think algorithms and with the ability to apply math concepts to be developed more efficiently (4) understand the consequences of scale, not only for efficiency reasons but also for economic and social reasons

According to Denning [4] the elements of computational thinking include the seven principles of calculation, communication, coordination, recollection, automation, evaluation, and design. These seven principles form a useful foundation for recognizing, organizing, and categorizing examples of computational thinking and building a framework that can translate computational thinking to contexts outside of computer science [5] Figure 1 shows the seven principles.

The seven principles mentioned above lead a meaning that CT is understandable by everyone [6], which means: (1) understanding which aspects of the problem are acceptable for computing, (2) evaluation of compatibility between tools and computing techniques and problems, (3) understanding the limitations and strengths of computing tools and techniques, (4) applying or adapting a computing tool or technique for new uses, (5) recognizing the opportunities to use computing in new ways (6) implementing computing strategies such as dividing and conquering in any domain.

3.2 Characteristics of Computational Thinking (CT)

As aforementioned, the article written by Wing [1] stated that CT has the following characteristics: (1) conceptualization, not programming (2) basic skills, not memorizing (3) the human way, not the computer (4) complete and combine mathematical thinking and techniques (5) ideas are not artefacts (6) for everyone everywhere (7) a scientific problem that is intellectually challenging and interesting (8) a person can master computer science and do anything

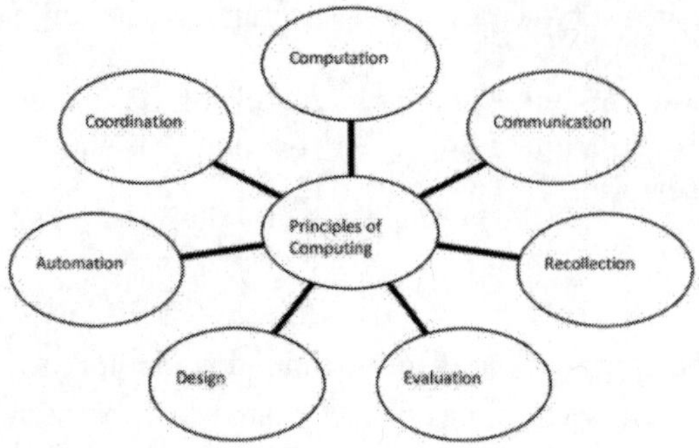

Figure 1. Seven principles of computational thinking.

CT is related to the ability to think recursively, using abstraction and decomposition when completing complex tasks. Some researchers focus on skills that are characteristic of CT. CT represents a cognitive process that accommodates logic, algorithmic, analytical, mathematical, technical and creative thinking [7] Three dimensions of computational thinking are computational concepts, computational practices and computational perspectives. The characteristics of computational thinking are being able to [1]: (1)provide troubleshooting using a computer or other device (2)organize and analyse data (3) perform data representation through abstraction with a model or simulation (4) automate solutions through algorithmic thinking (5) identify, analyse and implement solutions with various combinations of steps / ways and resources that are efficient and effective. (6) generalize solutions for different problems.

Thinking computationally is a fundamental skill for everyone, not just computer scientists. This suggests that the importance of integrating computing ideas into other disciplines. The following are elements or elements of CT. (1) **Decomposition**: The ability to break down data, processes or problems (complex) into smaller parts or become manageable tasks. For example, breaking up 'Drive / Direction' in a computer based on its constituent components: Files and Directory. (2) **Pattern Recognition**: The ability to see the similarities or even differences in patterns, trends and regularities in the data that will later be used in making predictions and presenting the data. For example, recognize the pattern of document files, file systems, execution files or data structures / files. (3) **Abstraction**: Generalize and identify general principles that produce such patterns, trends and regularities. For example, by placing all system files in the Windows folder, program files in the Program Files folder, data files / documents in the MyDocument Folder and supporting files in a separate Drive / Direction. (4) **Algorithm**: Develop the same step-by-step troubleshooting instructions so that others can use that step / information to solve the same problem.

3.3 Implementation of CT in ICT learning

In most societies today, teacher organizations, accrediting bodies, and government agencies are actively promoting the teaching of computational thinking (CT) skills [8] Discourse on the essence of CT continues to be done, integrated efforts have made computer science used as a requirement of high school graduation in many US states. Computational thinking (CT) is in the realm of educational innovation, as a part of problem-solving skills that students must acquire to thrive in a digital world full of software-driven objects. [9]

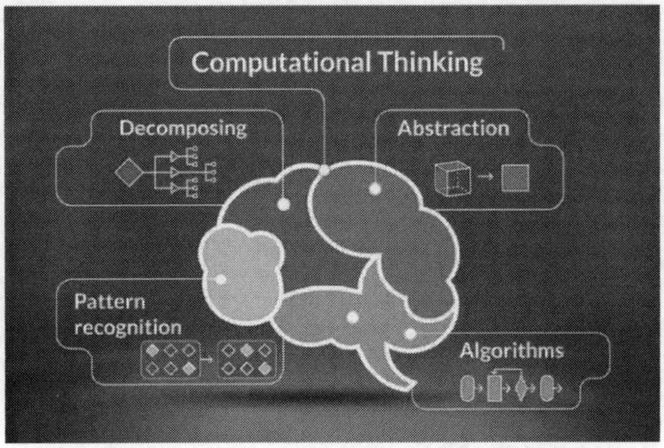

Figure 2. The elements of computational thinking.

Studies have been conducted to apply CT in schools, as has been done in K-12 [10] The study was to design standards of computational thinking curricula for K-12 education. The Delphi technique is used to collect different opinions and get agreement from thirteen expert discussions, including computer scientists, computer science educators, computer teachers and industry experts. The first draft of the Delphi questionnaire survey consisted of several themes: problem solving, problem decomposition, algorithm, data representation, data analysis, modelling and simulation, abstraction, and automation and then developed CT competence indicator based on the research. After three rounds of surveys and round-table meetings, experts produced 49 essential competency indicators (13 for grades 1 through 6, 9 for grades 7 through 9 and 27 for grades 10 through 12) and 8 optional competency indicators (1 for grade 7 up to 9 and 7 for classes 10 to 12). The result is that the capabilities and training of CT differ in each level.

In the Computational Thinking Pedagogical Framework (CTPF) associated with the pedagogical framework includes four pedagogical experiences one of which is Tinkering (tweaking) [11] The experience of fiddling primarily separates and engages in changes and / or modifications to existing objects. These objects can be building blocks, puzzles, digital or electronic simulations, programming codes, and so on. During the tinkering, students do not build objects, digital or otherwise, but rather explore changes to existing objects and then consider the implications of change. These experiences may require students to use some of the basic concepts and skills learned, but new concepts and skills that are being introduced can also be generated

In recent years, the availability of a free and user-friendly programming language has increased the interest of researchers and educators to explore how computational thinking can be introduced. Programming language is known as something hard to learn even many who believe only smart people who can operate it. However, the assumption is no longer appropriate, because now has developed a relatively easy to learn programming language that is Scratch programming [12] Scratch is a new programming language released in May 2007. Scratch supports the development of computer games, interactive stories, graphics and computer animation and other multimedia projects. One of the programming languages that uses the Tinkering method is Scratch. Below is an example of an overview of the use of the Scratch programming language.

Scratch programming does not use text form but in graphical form to put the codes used. Graphic forms are made in such a way that they form like puzzle pieces known as code blocks. To create a

IOP Conf. Series: Materials Science and Engineering **535** (2019) 012004 doi:10.1088/1757-899X/535/1/012004

Scratch program is like composing a puzzle. Code blocks can only be arranged if they can be interconnected. Thus the program created is protected from syntax errors. In the Scratch there are three types of code blocks and each can be a relation. The code block provided by Scratch 100 is further inserted into the 8 categories of code blocks.

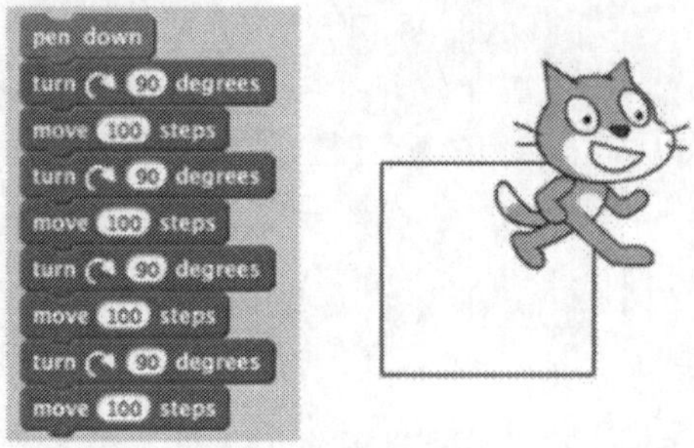

Figure 3. Draw a quadrilateral using Scratch.

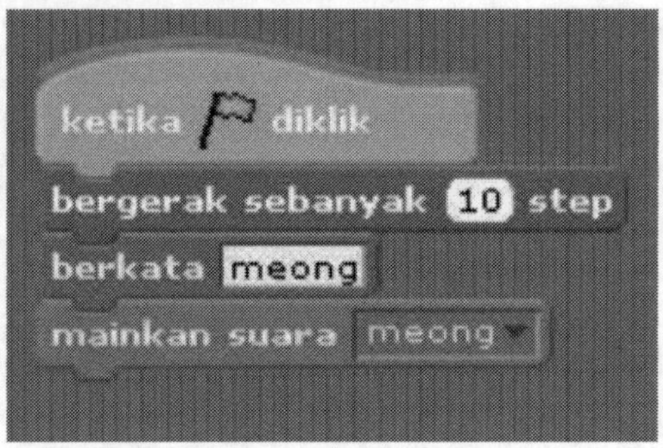

Figure 4. Block the script used as the base of script writing.

To create a Scratch app, required an object known as a sprite. A Scratch app can consist of many sprites. Each Sprite has several scripts, costumes and sound types. Scripts are programs for creating applications. The costume is the image that will be displayed. Medium types of sounds are the sounds that will be played in the app. The program results will be shown in an area called stage / stage.

In storing Scratch data using variable code block. Just like any other programming variables in Scratch can be created as local variables or global variables. This variable code block can be used to control application execution. The value of the variable code block can be displayed on the stage / stage. If no longer needed, the variable block code can be deleted. As for numerical manipulation, Scratch has a block of numeric codes. In this block of code there are several blocks of code for mathematical operations, be it arithmetic or logic. To support the in-app process, Scratch provides the control blocks used for looping and conditional processes. It can also be used to control and coordinate scripting activities by sending and receiving messages between sprites.

Learning using the Scratch programming language will make it easier for students to understand the exact programming concepts and ensure that novice programmers learn the right way to structure and formulate programming logic. Because Scratch is not a text-based programming language it is not necessary to follow a complex set of syntax rules, because misconduct that does not match the syntax rules of the programming language will cause the application to not run. Scratch simplifies application development using the same basic logic and programming concepts with other programming languages

To measure the extent of CT capability use measurement scale, developed for the purpose of determining the level of computational thinking ability (CTS) of students [13] The CTS used is a five-point Likert scale and consists of 29 items from 5 developed indicators. The validity and reliability of the scale has been studied by exploratory factor analysis, confirmatory factor analysis, goods specificity analysis, internal consistency coefficient and analysis of firmness. As a result of the analysis, it has been concluded that the measurement scale is valid and reliable as a tool that can measure students' computational thinking skills. However, the validity and reliability of such scales can only be used in certain groups. If it is to be used in different groups, renewals should be made in accordance with the group.

4. Conclusion

Computational Thinking (CT) is basically problem solving, designing systems, and understanding human behaviour, by drawing on fundamental concepts to computer science which can be done by humans or machines (computers). The goal of embedding computational thinking skills is to prepare students with the ability to produce computational solutions to problem solving. CT has strong principles and characteristics that serve as a basis for CT development and its implementation at several levels of education that have applied CT a lot in the curriculum.

Several levels of education have implemented a CT-based curriculum and have undertaken research that supports the further development of CT. Implementation of CT at the level of education is completed to facilitate students to solve problems using computers with programming language that is easy to understand and use. Then some research has been conducted to develop the scale of assessment to measure the level of ability of CT in students. The scale is considered valid for certain groups only.

Further CT development is CT can be the basis for the development of ICT curriculum nationally, because at this time ICT learning using different curriculum concepts of guidance. It thus must be made a curriculum design that is packed in such a way for the more structured ICT counselling process.

5. References

[1] J. Wing, *Computational Thinking*, Communications of the ACM 49(3), pp. 33-35, 2006.

[2] Sze Yee Lye , Joyce Hwee Ling Koh, *Review on teaching and learning of computational thinking through programming: What is next for K-12?* National Institute of Education, Nanyang Technological University, Singapore, 1 Nanyang Walk, Singapore 637616, Singapore Computers in Human Behavior 41 (2014) 51–61.

[3] James J. Lu, George H. L. Fletcher, *Thinking About Computational Thinking, SIGCSE'09,* March 3–7, 2009, Chattanooga, Tennessee, USA, Copyright 2009 ACM 978-1-60558-183-5/09/03.

[4] Denning Peter J, *Great principles of computing*, Communications of the ACM-Blueprint for the future of high-performance networking, Volume 46 Issue 11, November 2003, Pages 15-20.

[5] Perkovi´c Ljubomir, Settle Amber, *Computational thinking across the curriculum: A conceptual Framework,* 2009. http://compthink.cs.depaul.edu/ FinalFramework.pdf, Accessed October 12, 2015.

[6] Cuny Jan, Snyder Larry and Wing Jeannette M, *Demystifying Computational Thinking for Non-Computer Scientists*, work in progress, 2010.

[7] Ana Liz Souto O. de Araujo, Wilkerson L. Andrade, Dalton D. Serey Guerrero, *A Systematic Mapping Study on Assessing Computational Thinking Abilities,* Software Practices Laboratory Federal University of Campina Grande Departament of Exact Sciences Federal University of Para′ ıba Para′ ıba, Brazil. Wing.

[8] Osman Yasar, *The Essence of Computational Thinking*, State University of New York, Brockport.

[9] Marcos Roman-Gonzalez, Juan-Carlos Perez-Gonzalez, Carmen Jimenez-Fernandez, *Which cognitive abilities underlie computational thinking? Criterion validity of the Computational Thinking Test,* Computers in Human Behavior xxx (2016) 1e14.

[10] Hui-Chi, Chuang; Chiu-Fan, Hu; Cheng-Chih, Wu; Yu-Tzu, Lin, *Computational Thinking Curriculum for K-12 Education-A Delphi Survey,* 2015 International Conference on Learning and Teaching in Computing and Engineering.

[11] Donna Kotsopoulos & Lisa Floyd & Steven Khan & Immaculate Kizito Namukasa & Sowmya Somanath & Jessica Weber6 & Chris Yiu, *A Pedagogical Framework for Computational Thinking.*

[12] Ekohariadi, 2015, Ibnu Febry Kurniawan, Ricky Eka Putra, Pemrograman Visual Menggunakan Scratch, Unesa University Press

[13] Özgen KORKMAZ *TURKEY* Recep ÇAKIR, *A validity and reliability study of the Computational Thinking Scales (CTS).*

IOP Conf. Series: Materials Science and Engineering **535** (2019) 012005 doi:10.1088/1757-899X/535/1/012005

Development of critical and creative thinking skills to increase competence of PLC programming for electrical engineering education students

PW Rusimamto[1], L Nurlaela[2], MS Sumbawati[3], Munoto[4] and M. Samani[5].

[1,2,3,4,5]Universitas Negeri Surabaya, Surabaya, Indonesia

Email: puputwanarti@unesa.ac.id

Abstract. PLC learning emphasizes on teaching and learning activities as well as developing the concepts and skills of the students process with various teaching methods that are in accordance with the study materials being taught. PLC learning demands an active role for students, because PLC is a scientific process based on logical thinking based on supporting facts. The low ability of students thinking is caused by PLC learning so far only tends to hone aspects of remembering and understanding. Basically students have critical thinking skills in learning such as questioning skills, hypothesis, classification, observation and interpretation. Problem-based learning model (PBL) is a learning model designed on problem solving process that faced scientifically so that students get important knowledge. Thus expected students are adept at solving problems, have their own learning models and have the skills to participate in the team. Teacher's role as a trainer in problem-based learning is asking about thinking, monitoring learning, probbing (challenging students to think), keeping students engaged, organizing group dynamics, keeping the process going. This critical and creative thinking design aims to make students able to think critically and able to consider everything and be able to act on the decisions that have been considered and creative thinking that aims to learners are able to create ideas, or ideas and able to express the idea to solve a problem.

1. Introduction

Programmable Logic Controller (PLC) is a compulsory subject for the undergraduate students of Electrical Engineering Education. One of the goals of Electrical Engineering Education Unesa S1 is to produce competent teachers in electrical engineering. Among them as subject teachers on the competence of instrumentation and process automation skills on the spectrum of SMK 2016 and in accordance with the circular of the Director of Vocational Education on the Implementation of Vocational High School Curriculum Number 4540 / D5.3 / TU / 2017, dated June 22, 2017[1].

PLC learning emphasizes on teaching and learning activities as well as developing the concepts and skills of the students process with various teaching methods that are in accordance with the study materials being taught. PLC learning demands an active role for students, because PLC is a scientific process based on logical thinking based on supporting facts. The low ability of students thinking is caused by PLC learning so far only tends to hone aspects of remembering and understanding. Basically students have critical thinking skills in learning such as questioning skills, hypothesis, classification, observation and interpretation. However, these skills sometimes do not develop well, then there is an alternative method that is able to develop students' critical thinking skills in PLC learning. One of them is through practicum or demonstration activities, because practicum or demonstration activities help students to understand an event, see an event more detailed than before and then remember the incident.

Wechsler (2017) says that creative or critical thinking is needed to develop problem-solving skills[2]. The results of his research say that there is differentiation and independence of creativity and critical thinking in cognitive performance, although both build small but complementary roles in different creative problem solving phases. As observed, the flexibility, ie the ability to consider the same information from different angles (Torrance, 1990), can improve critical thinking skills[3]. The ability of flexibility and critical thinking is considered a key ingredient to promote examination and resolution of problems in many professions (Hong & Choi, 2015)[4].

On the other hand, the creative process supports inductive reasoning, which is the ability to understand relationships. The importance of making connections through analogies and metaphors is an important characteristic of creative thinking (Wechsler, 2006) but intellectual critical functions can also contribute to this process[5]. With this in mind, creativity prefers inductive reasoning, especially in the identification of order and principle or conclusions about relationships, which do not occur in relation to logical reasoning, focusing on finding the best unique answer.

Dwyer (2016) says that critical thinking (CT) is a metacognitive process, consisting of a number of subclasses and dispositions which, when used appropriately, increase the chances of generating logical conclusions to the argument or solution to a problem[6]. There is potential to enhance the development of critical thinking by assessing effective writing with CACTiM tools (reflection) to provide formative and summative feedback to students and inform instructional strategies disclosed by Carter (2016)[7]. From Okan research (2012) stated that students' critical thinking skills vary between types of schools[8]. When students encounter situations or problems they are unable to cope with, to get help, they must communicate with their classroom teachers in primary school and with their counselors in high school. Thus, teachers should be prepared to perform services to help students optimally and approach the problem logically.

From these reviews, then made the problem formulation is how to develop the ability of critical and creative thinking? And how to improve the competence of PLC programming students Electrical Engineering Education?

2. Library Review

2.1. Critical Thinking

Thinking is a mental activity that a person experiences when they are faced with a problem or situation to be solved. Suryabrata (1990) argues that thinking is a dynamic process that can be described by process or path. The process of thinking that essentially consists of 3 steps, namely the formation of understanding, formation of opinions, and conclusions[9]. This view indicates that if a person is faced with a situation, then in thinking, the person will establish the relationship between the parts of the information recorded as the notions. Then the person forms opinions that are in accordance with his knowledge. After that, he will make the conclusions that are used to discuss or find a solution of the situation.

Ruggiero (1998) defines thinking as a mental activity to help formulate or solve a problem, make a decision, or fulfill the curiosity desire. This suggests that when a person formulates a problem, solves a problem, or wants to understand something, he performs an activity of thinking.

Thinking as a person's mental ability can be divided into several types, including logical, analytical, systematic, critical, and creative thinking. Logical thinking can be interpreted as a student's thinking ability to draw legitimate conclusions according to the rules of logic and can prove that conclusions are valid in accordance with previously known knowledge. Analytical thinking is the thinking ability of students to describe, itemize, and analyze information used to understand a knowledge by using logical reason and thought, not based on feelings or guesswork. Systematic thinking is the ability to think students to do or complete a task in accordance with the sequence, steps, steps, or planning the right, effective, and efficient. The three types of thinking are interrelated. Someone to be said to think systematically, then he needs to think analytically to understand the information used. Then, to be able to think analytically necessary logical thinking ability in drawing conclusions to a situation.

Critical thinking and creative thinking is the embodiment of higher-order thinking. This is because the ability to think is the highest cognitive competence that students need to master in the classroom. Critical thinking can be viewed as a student's thinking ability to compare two or more information, eg information received from outside with information held. If there are differences or similarities, then he will ask questions or comments with the aim to get an explanation. Critical thinking is often associated with creative thinking.

Some of the thinking skills associated with critical thinking are comparing, differentiating, estimating, drawing conclusions, influencing, generalizing, specializing, classifying, grouping, sorting, predicting, validating, proving, linking, analyzing, evaluating and pattern-making. A student is said to be able to think critically if it has the ability to: 1) choose important words and phrases in a statement and will be carefully defined; 2) requires confidence to support a conclusion when he is forced to accept it; 3) (5) evaluating these assumptions, accepting only a few and rejecting others, 6) evaluating opinions, accepting or rejecting conclusions and (7) evaluating the assumptions, accepting or rejecting the conclusions; and 7) continually re-examining assumptions that have been made and believed previously.

Critical thinking processes include: 1) recognizing situations, 2) considering opinions according to evidence, data, or assumptions, 3) providing arguments beyond proof, 4) reporting and supporting conclusions / decisions / solutions, and 5) applying conclusions / decisions / solutions . One approach that can be used to increase students' sense of critical thinking is by creating a classroom atmosphere where students feel comfortable questioning something, challenging, suspending judgments, and demanding reason and justification because they are dealing with real-world contents and math. Ask questions that stimulate students to monitor, evaluate, and act on their own thinking. For example, Invite students to work in groups to (a) Discuss the situation below, (b) Brainstorming ideas to solve it. (c) Find solutions that all received, or this minority report, and (d) Discuss their thinking to arrive at a decision.

2.2. Creative Thinking

Creative thinking or creativity is often exchanged for its meaning. Weisberg (2006) defines creative thinking as referring to processes to produce a creative product that is a new work (innovative) derived from an activity / activity that is directed to the goal. Another sentence is said to think creatively involves intensive production that meets novelty, so that one can be said to be creative by producing something already known before. If you produce something new to you, but it has been generated by others, then you can still be said to be creative.

In looking at the link between creative thinking and critical thinking there are two views. The first view of creative thinking is intuitive in contrast to critical thinking based on logic, and secondly looking at creative thinking is a combination of analytical and intuitive thinking. Intuitive thinking means thinking of getting something by instinct or feeling suddenly without any general facts. The first view tends to be influenced by views of dichotomies of the right and left brain which have different functions, while the second view sees the two hemispheres working synergistically together that are not separated.

Creative thinking is a mental activity that considers authenticity and insight. Thinking critically and creatively allows students to study the problem systematically, bringing together many challenges in an organized way, formulating innovative questions and designing / designing original solutions. Creative thinking as opposed to destructive thinking involves searching for opportunities to change things for the better. Creative thinking does not explicitly organize processes, such as critical thinking. Creative thinking is a habit of sharp thinking with intuition, moving imagination, revealing new possibilities, unveiling amazing ideas and inspiring unexpected ideas. This understanding distinguishes firmly from creative thinking and critical thinking.

De Bono (in Barak and Doppelt, 2000) distinguishes between two types of thinking, ie lateral thinking and vertical thinking. Lateral thinking refers to the discovery of new clues in the search for ideas, while thinking vertically in the face of the development of ideas and examination of an objective criterion. Vertical thinking is selective and sequential that moves only when there is a clue in motion. Lateral thinking is a generative that can jump and move in order to build a new clue. Lateral thinking does not have to be true at every step and does not use category categorization, classification or fixed labels. Vertical thinking selects a very promising approach to a problem during lateral thinking building many alternative approaches. Creative thinking is a synthesis of complementary lateral and vertical thinking. This sense states that in creative thinking involves critical thinking (logical and analytical) as well as intuitive, as in the second view in terms of creative thinking.

2.3. Programmable Logic Controller

Programmable Logic Controller (PLC) is a computer specially designed for use on industrial machines. This computer has been designed for use in industrial environments, equipped with special input / output and a programming language for control.

2.3.1. Hardware

PLC is a type of control system that has input device, controller and output device. The equipment connected to the PLC which sends a signal to the PLC is called the input divice. While the controller is to perform the calculation, decision-making, control of the input to be issued in the output. How the PLC works can be explained through the block diagram in Figure 1 below. From Figure 1 shows that the PLC construction consists of CPU, input device, output device, supporting equipment, and power supply.

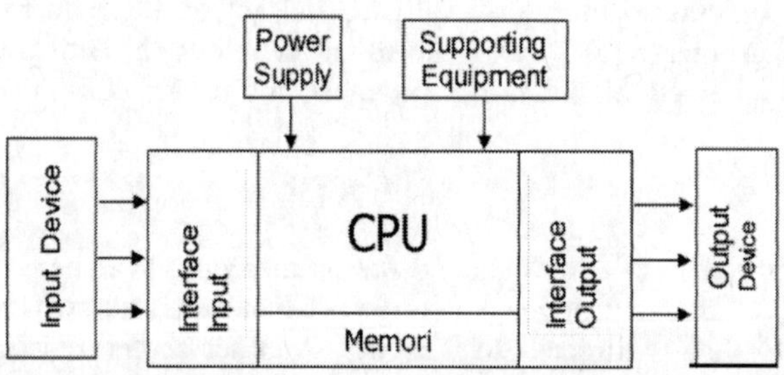

Figure 1. PLC block diagram

2.3.2. Basic Instruction

The basic instructions that exist in programming using PLC are LD (Load), AND, OR, OUT, AND NOT, OR NOT, TIMER (TIM), COUNTER (CNT).

2.3.3. Ladder Diagram and Mnemonic Code

To make it easy to write and insert program on PLC then needed some basic stage. Ladder diagram of a program is made first to facilitate the preparation of mnemonic code. The program form mnemonic code can be directly inserted into the CPU via a PC. The ladder diagram consists of a line extending downward from the left side with its branches towards the right. The line extending down on the left side is called the busbar. While the branches are called the instruction line. Along the lines of instruction are placed conditions leading another instruction line on the next right side. The logic combination of these conditions determines when and how the instructions on the right side are executed.

IOP Conf. Series: Materials Science and Engineering **535** (2019) 012005 doi:10.1088/1757-899X/535/1/012005

2.3.4. *Software*
There are five ways to program PLC: (1) Programming Console (PC), (2) Prompt Writer. (3) Sysmac Support Software + Personal Computer. (4) Factory Intelegent Terminal (FIT). (5) Graphic Programming Console.

3. Discussion

Science is a system of thinking about the empirical world. Therefore learning needs to develop rational thinking skills about the empirical world. From the taxonomic side of thinking, hence-learning means educating thinking at a certain cognitive level. With Bloom's taxonomy (2002), for example, critical and creative thinking lies at the level of analysis-synthesis-evaluation-creation, not at the lower levels of remembering, understanding, and applying[10]. When using the Merrillian taxonomy (1983), the thought lies at the level of finding, not at the lower level of remembering and using. In PLC learning where required to be able to do programming with the help of software through personal computer and or with programmable console that requires thinking not only remember it but must use logic through the knowledge of logic circuit[11]. Competencies to be met in PLC learning are hardware, basic instructions (LOAD, AND, OR, OUT, AND NOT, OR NOT, TIMER, COUNTER), ladder diagrams, mnemonic code, and software. To improve the PLC programming competence, it is necessary to improve the learning process through the application of variation of learning model and the use of appropriate learning media and cover affective, cognitive and psychomotor spheres. The ability to think critically and creatively to improve the competence of PLC programming is used problem-based learning model. Problem-Based Learning Model (PBL) is a learning model designed on problem solving process that is faced scientifically so that students get important knowledge. Thus expected students are adept at solving problems, have their own learning models and have the skills to participate in the team. Teacher's role as a trainer in problem-based learning is asking about thinking, monitoring learning, probbing (challenging students to think), keeping students engaged, organizing group dynamics, keeping the process going. Student roles as problem solver in problem based learning are active participants, directly involved in learning, build learning.

Generally problem-oriented learning can improve the ability of critical thinking and creative thinking. Such learning is like problem-solving, problem-solving, or problem-based learning. To judge creative thinking actually depends on the criteria or indicator of creative thinking formulated by the researcher or its users. Plucker, Beghetto, & Dow in Plucker and Makel (2010) recommend that "all examinations of creativity clearly define the authors' conception of creativity as used in that work"[12]. Guilford, Torrance, Wallach and Kogan, Getzels and Jackson use divergent thinking criteria to judge a person's creative thinking. While using varying content and instructions, diverging thinking equals multiple responses and is assessed using fluency criteria, flexibility, authenticity, and elaboration of ideas. According Dimyati (1996) one of the elements of science is the items, namely the science of tangible rational thinking[13]. The realization of rational thinking appears in the use of words, sentences, aleneas, problem solving formulas, or symbols. Prerequisites for realizing these items are the individual's ability to read, write, think and observe. In other words the requirement is the ability to think critically and creatively. The learning process of PLC with PBL model applied to the student of Electrical Engineering Education program class of 2015 cultivates creativity and critical thinking when solving the given problem shown in Figure 2.

Figure 2. Students are solving problems using a PLC training kit at the Unesa Electrical Engineering Control Laboratory.

With PLC media training kits and various problems given, it shows the learning outcomes with an average score of 73.83 with a scale of 100. Learning outcomes are shown in Table 1 below. The lowest score is 55.5 and the highest score is 85. Students whose value is above average is 21 and below the average is 15 out of a total of 36 students.

Table 1. Results of PLC courses in Electrical Engineering.

No	Student Number	Learning outcomes
1	15050514002	72.5
2	15050514006	80
3	15050514007	77.5
4	15050514008	77.5
5	15050514009	60
6	15050514011	75
7	15050514013	55.5
8	15050514015	77.5
9	15050514022	70
10	15050514023	77.5
11	15050514028	70
12	15050514030	75
13	15050514031	80
14	15050514034	77.5
15	15050514036	77.5
16	15050514038	77.5
17	15050514039	72.5
18	15050514041	80
19	15050514042	77.5
20	15050514043	80
21	15050514045	67.5
22	15050514050	77.5
23	15050514052	77.5
24	15050514053	72.5
25	15050514054	82.5
26	15050514056	75
27	15050514057	77.5
28	15050514060	72.5
29	15050514063	67.5
30	15050514066	67.5
31	15050514067	85
32	15050514070	72.5
33	15050514072	75
34	15050514073	72.5
35	15050514074	67.5
36	15050514078	57.5

IOP Conf. Series: Materials Science and Engineering **535** (2019) 012005 doi:10.1088/1757-899X/535/1/012005

4. Conclusion

Critical and creative thinking are two inseparable sides and educational goals everywhere. Both are necessary skills in everyday life. To be applied depends on the ability and confidence of the lecturer to apply in the form lab in the laboratory. With the help of training kits, students will be more optimal in understanding the material being taught. Does not depend on difficult or difficult material. Simple problems can lead students to achieve that goal. The learning culture in the classroom that needs to be changed by giving students flexibility to argue and argue can create a critical attitude. By giving the opportunity to give other ideas or other strategies though not the same from the habit can create creative students. Then appreciate every wrong or true task result can make learning fun.

To improve the competence of PLC programming students Electrical Engineering Education, teachers apply problem-based learning model (Problem Based Learning). One of the learning design in order to make the atmosphere of learning activities fun and produce an effective and productive learning process that is critical and creative thinking. This critical and creative thinking design aims to make students able to think critically and able to consider everything and be able to act on the decisions that have been considered and creative thinking that aims to learners are able to create ideas, or ideas and able to express the idea to solve a problem.

5. References

[1] Surat Edaran Direktur Pembinaan SMK Tentang Pelaksanaan Kurikulum Pendidikan Menengah Kejuruan Nomor 4540/D5.3/TU/2017, tanggal 22 Juni 2017.

[2] Wechsler, Solange Muglia., Saiz, Carlos., Rivas, Silvia F., Vendramini, Claudete Maria Medeiros., Almeida, Leandro S., Mundim, Maria Celia., & Franco, Amanda. (2017). Creative and critical thinking: Independent or overlapping components?. *Thinking Skills and Creativity* https://doi.org/10.1016/j.tsc.2017.12.003.

[3] Torrance, E. P. (1990). *Torrance Tests of Creative Thinking: Figural and Verbal A*. Benseville, IL: Scholastic Testing Service.

[4] Hong, Y. C., & Choi, I. (2015). Assessing reflective thinking in solving design problems: The development of a questionnaire. *British Journal of Educational Technology*, 46(4), 848-863. doi: 10.1111/bjet.12181.

[5] Wechsler, S. M. (2006). Validity of the Torrance Tests of Creative Thinking to the Brazilian culture. *Creativity Research Journal*, 18(1), 15-25. DOI: 10.1207/s15326934crj1801_3

[6] Dwyer, Christopher P., Hogan, Michael J., Harney, Owen M., Kavanagh, Caroline. (2016). Facilitating a student-educator conceptual model of dispositions towards critical thinking through interactive management. *Education Tech Research Dev* (2017) 65:47–73. DOI 10.1007/s11423-016-9460-7.

[7] Carter, Amanda G., Creedy, Debra K., Sidebotham, Mary. (2017). Critical thinking evaluation in reflective writing: Development and testing of Carter Assessment of Critical Thinking in Midwifery (Reflection). *Midwifery* 54 (2017) 73–80. DOI 10.1016/j.midw.2017.08.003.

[8] Sarigoz, Okan. (2012). Assessment of the high school students critical thinking skills. *Social and Behavioral Sciences* 46 (2012) 5315 – 5319. DOI 10.1016/j.sbspro.2012.06.430.

[9] Suryabrata, Sumadi. (1990). *Psikologi Pendidikan*. Jakarta: CV Rajawali.

[10] Bloom William, Don W. Fawcett. (2002). *Buku ajar histologi*. Edisi 12. Terjemahan Jan Tambayong. Jakarta: EGC.

[11] Merrill, D. (1983). Component Display Theory. In C. M. Reigeluth (ed), Instructional Design Theories and Models: *An Overview of their Current States*. Hillsdale, NJ: Lawrence Erlbaum.

[12] Plucker, Jonathan A. and Makel, Mathew C. (2010). Asssessment of Creativity. In Kaufman, James C & Sternberg, Robert J (Eds). *The Cambridge Handbook of Creativity*. Cambrigde: Cambridge University Press.

[13] Dimyati. (1996). Guruan Keilmuan di Indonesia: Suatu, Dilema Pengajaran dan Penelitian. Jurnal *Guruan Humaniora dan Sains*-.September.

IOP Conf. Series: Materials Science and Engineering **535** (2019) 012006 doi:10.1088/1757-899X/535/1/012006

Analysis of an industrial internship in the department of mechanical engineering with servqual, kano, and quality function deployment

FA Devi[1], Z Rofiq[2] and E Purnomo[3]

[1,2,3]Department of Mechanical Engineering Education. Yogyakarta State University, Yogyakarta, Indonesia

E-mail: fauziahamalia@uny.ac.id

Abstract. *The objective of this study is to examine: (1) the real implementation of an industrial internship in Vocational High School 2 Klaten (2 the level of gap between the perceived quality and the expected quality of the implementation (3)* attribute importance *needed in improving the quality of the implementation (4) priorities that the school management must do for improving the quality. Data analysis used servqual, Kano model and quality function deployment. The data collection techniques were observation, questionnaires, interview, and documentation. The results of this study indicated the average performance based on the rating of students, teachers, and industry under expectation score (priority score). The gap level of tangibles was -0,47, reliability was -0,51, responsiveness was -0,57, assurance is -0,44 and empathy is -0,20. All of the attributes are important to improve the quality of industrial internship program. Prioritized strategies had to be done by school's management were: (1) preparing infrastructure for internship participants (2) determining the feasibility standards of companies involved in the internship program (3) moral development of internship participants (4) determining duties and obligation of the supervisor teachers (5) setting the standard operating procedure of pre-implementation of the program (6) choosing standard activities for internship participants (7)setting the policy of internship's wearpack*

1. Introduction

The social paradigm of vocational education development as economic since vocational education has a significant role in developing modern industry and improving the quality of life of the community. [1] states that vocational education can reduce the rate of unemployment. This is in line with the theory that a Vocational High School is one of the formal education institutions to produce workers equipped with skills and expertise in accordance with the needs of the young generation to obtain employment opportunities for national development purposes. Cooperation between VHS with the world of work is absolutely necessary to meet the demands of learning and effectiveness in producing skillful graduates according to community expectations [2]. This is in accordance with the study presented by [3] which stated that:

With globalization and the liberalization of higher education in Malaysia, universities and colleges face new challenges. Given this head-on competition, universities and colleges have no choice but to improve the quality of their programs and graduates so that they can compete globally and attract students. Higher education is increasingly recognized as a service industry, placing greater emphasis on meeting the expectations and needs of its participating customers namely the students

It is also supported by a study conducted by [4], which concludes that the quality of service in education and training should get full attention. To conclude, in the era of globalization, the secondary schools have no other choice but to develop the quality of the program. This is very important because students must be able to compete in various industry sectors .Information obtained from pre-survey results in BKK, Purwoko and the team from BKK stated that there are problems regarding the quality of Industrial Internship implementation in VHS 2 Klaten. These problems illustrate poor quality in the implementation of Industrial Internship. In facts, according to [5] in the implementation of education, needs and ideas of students should be the main focus. Students are the main reason for the establishment of an educational institution and the reputation of educational institutions is on the students' shoulders.

Drs. Purwoko as the chairman of the program said that the implementation of Industrial Internship from year to year was no effectively assessed. This fact shows that the quality of service in the implementation of the Industrial Internship is still neglected. Quality assessment of industrial practice will result in outcomes in the form of customer satisfaction levels of service. Consumer satisfaction can be achieved by providing good quality. Customer satisfaction results from an evaluation process that compares the expectations of customers with the actual situation regarding the implementation of the Industrial Internship program at VHS 2 Klaten. Considering the importance of service quality for education services business, [6] claimed that it takes hard efforts in order to remain competitive and high quality. With such conditions, it is not enough if the educational institutions depend only on the continuous improvement to maintain and develop their competitiveness. They have to foster innovation. Therefore, a quality service study usually performed with Servqual is required. In this study, a quality study of Industrial Internship implementation by integrating Servqual with Kano Model into Quality Function Deployment (QFD). According to Tan and Pawitra in [6], service quality measurement with servqual should be followed by application of QFD to clarify the action plan that must be done to close the gap. Quality of service is formed by the comparison between the ideal with the perception of the quality dimension performance.

The advantage of using these three methods is to focus on improvements that should be made to improve the implementation of industrial practices in VHS 2 Klaten. [7]. This integration aims to assist BKK of VHS 2 Klaten to assess the process of the implementation of Industrial Internship, not only to assess consumer expectations have been met or not but also accelerate the development of service innovation by identifying the attributes and applying them to customer satisfaction. In addition, VHS 2 Klaten is one of the vocational schools with an advanced level of information technology and teaching and learning facilities which should be able to do continuing development and improvement. The identified problems are: there is no assessment of the real condition of the implementation of industrial internship in VHS 2 Klaten, there is no study on the gap between the perceived quality and the expected quality of implementation of the Industrial Internship, there is no data of urgent concerns in improving the quality of the implementation of Industrial Internship, there is no data on priorities to be set by the school management as an effort to improve the quality of the implementation of Industrial Internship. The study was conducted in the even semester of the academic year of 2016/ 2017 on the students of XIII grade of Mechanical Engineering Department. The data would be used was primary data obtained directly from respondents in the field by directly distributing questionnaires. On Servqual measurement, it was only measured gap 5 that was the gap between the perceived service and expected service by the service users. The QFD stage used consisted of two stages. The student behavioral factors are considered normal based on the information from the school.

This study was aimed to determine the real condition of Industrial Internship implementation in VHS 2 Klaten, the level of gap between the perceived quality and the expected quality of service in the implementation of Industrial Internship, attribute importance in improving the quality, priorities to be conducted by the school as an effort to improve the quality.

2. Method

This study was conducted through a descriptive approach with a combination of quantitative and qualitative studies. The descriptive approach according to philanthropist [8] has the meaning of a study that refers to the transformation of raw data into a form that is easily understood and translated. The study was conducted in Klaten Regency at VHS 2 Klaten.

The population in this study consisted of 45 students of VHS 2 Klaten in the Program of Mechanical Engineering, 20 involved industries from Klaten Regency, 26 teachers assigned as BKK managers in VHS 2 Klaten.To determine the sample in this study the researcher refers to the table of the number of samples based on krejcie and morgan tables. Therefore, it was decided the number of students in VHS 2 Klaten was 45 students, the industries involved in Industrial Internship program in Klaten Regency were 20 industries, the managers of BKK in VHS 2 Klaten is 26 teachers.

2.1 Data collection techniques

The first data collection technique was observation. The second technique was administrating questionnaires. The third technique was an interview. The last was documentation which was conducted by collecting secondary data in the form of documents of industrial practice activities.

2.2 Research Instruments

Questionnaires were required to obtain research data. The research instrument is a measuring instrument used by researchers in collecting data. According to [7], the steps of combining Servqual and Kano Models, and the integration into House of Quality in QFD are presented in Figure 1

Table 1. Research Instruments

Indicators	Aspects	Item
Tangible	Students	37,53
	Supervisor and the school	1,2,3,4,5,6,7,8,10,38,54
	Industry	9,39,40,41
Reliability	Students	56,57,58
	Supervisor and the school	11,12,13,14,15,44,55
	Industry	42,43,45,59
Responsiveness	Students	60,61,62
	Supervisor and the school	16,17,18,19,20,21,22
	Industry	46,47,48,49
Assurance	Students	64, 65
	Supervisor and the school	23,24,25,26,27,28,63
	Industry	29,30,31,50,51
Empathy	Students	67,68,69
	Supervisor and the school	32,33,34,35,36,66
	Industry	35,52

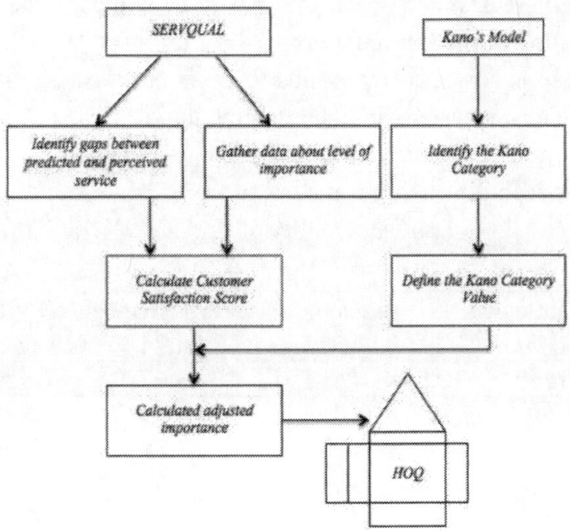

Figure 1. Work Framework in Combining Servqual Method, Kano Method, and QFD

3. Result and Discussion

Validity tests include content validity and construct validity. The instrument used in this study had been validated by expert judgment. The validity of the constructs was measured using Bivariate Pearson. All grains were declared valid. Reliability testing was used reliability formula by Alpha coefficient formula Reliability is to test the reliability of the items that are considered valid or invalid variables [9]. The result of the reliability test is reliable if the coefficient value of alpha obtained was bigger than the table with 5% significance. The results showed that all instruments in this study were reliable. Based on the results of interviews and observations on the students, schools, and industries, it can be identified 69 attributes of Industrial Internship services that become Customer requirements in the implementation of industrial practices. Identification is based on the 5 components of service quality, namely: Tangibles, Reliability, Responsiveness, Assurance, and Empathy.

Table 2. Expectation, Perfomance and Gap of Each Dimension

No	Dimension	Expectation	Performance	Gap
1	*Tangible*	4,213	3,74	-0,47
2	*Reliability*	4,18	3,67	-0,51
3	*Responsiveness*	4,16	3,59	-0,57
4	*Assurance*	4,32	3,88	-0,44
5	*Empathy*	4,26	4,06	-0,20
	Mean	4,228	3,79	-0,44

The obtained data indicated that the gap score obtained by service attribute in industrial practice is <0. It means performance attribute of service of industry practice was still below the expectation or the expectation value. It can be said that the implementation of the Industrial Internship did not satisfy the subjects of the implementation. This is because the service can be called satisfactory is if there is a positive gap score, whereas if the negative score gap means the quality of service cannot satisfy the service users. Whereas if the score the gap is 0, it means the quality of service is in accordance with the expectation of the service users.

Although the negative gap is small, or the minus value is not too large it is still not satisfactory. Servqual adopts a zero-tolerance system in its implementation. So, all attributes are still below satisfactory quality. Based on the expectations and judgments obtained for each dimension, then the classification of qualities for each dimension can be seen from Table 3.

Table 3. Quality of Service

No	Dimension	Expectation	Performance	Q = P/E
1	*Tangible*	4,213	3,74	0,89
2	*Reliability*	4,181	3,67	0,88
3	*Responsiveness*	4,163	3,59	0,86
4	*Assurance*	4,320	3,88	0,90
5	*Empathy*	4,262	4,06	0,95
	Mean	4,228	3,79	0,90

Based on the calculation of the quality of each dimension in Table 3, the quality is categorized as good if (Q) ≥ 1 then the service implementation of the Industrial Internship was still not good. Therefore it is necessary to improve the service attributes. In doing the improvement, it is better to prioritize attributes that can increase customer satisfaction. Therefore, the next stage attributes of industrial practice service quality will be analyzed by a Kano model.

In the analysis using the Kano model, there is a classification of service attributes, because basically, Kano Model is to do the classification by looking at the relationship between the level of customer satisfaction over service attributes and the degree of fulfillment of the service attribute itself. By classifying the service attributes, the service attributes can be divided into 6 categories namely A (Attractive), O (One Dimensional), M (Must Be), I (Indifferent), Q (Questionable), and R (reverse).The results of the Kano data processing showed that there were several attribute categories that emerged, among others: there were 43 attributes that belong to category O, there were 20 attributes that must be (M) category, there were 2 attributes that enter category A (attractive), there were 2 category I (Indifferent). Based on the Kano analysis results, there were two indifferent variables (their existence is not important) namely variables 44 and 54 that are conformity of materials delivered in the school with works in industrial internship and the existence of MoU between the industry and the school. The emergence of two variables that enter into this indifferent category can be caused by several factors among others: respondents did not understand the advantages and disadvantages of the MoU in the implementation of Industrial Internship. When filling the questionnaires respondents consider that MoU is not significant and conformity between materials delivered in the school and industry is not important. According to the interview, the industry is not too concerned with the hard skills of students but more on soft skills. The teachers of VHS 2 Klaten had the same opinion in this case.

Therefore, these variables were not included in data processing using the Quality Function Deployment method because the existence of these attributes would not increase the satisfaction of stakeholders of the Industrial Internship. The integration of Service Quality method with the Kano method was to complement each other and cover the weakness owned by each method. Kano model covered the weakness of existing linear assumption on servqual method, while on the other hand, servqual could give an explanation about the attributes performance and the gap score of each of these attributes.

For example, using Kano methods, it can be seen that it is not fully assumed that the level of customer satisfaction will be linear with the level of fulfillment of service attributes that become consumer needs as analyzed in the servqual method. This assumption by Kano is not entirely correct because the assumption applies only to attributes of category O. On the other hand, the Kano model only focuses on the classification of products or service attributes. In the Kano model, it is not given a clear picture of the performance of the attributes and we will not find a gap score as the description. By using a servqual method, it can be seen the performance of 69 service attributes that exist with the gap score obtained. With the Kano model, we can classify which attributes should be developed and what strategy should be done based on those results, the service attribute that belongs to category A must be developed as this attribute is an innovative process. While attributes that have a positive score and category M should be used as a strength, therefore, the quality of these attributes must be maintained in accordance with consumer expectations. This is because it is a basic attribute that must meet customer expectations, as well as service attributes that category O should be developed and maintained.

Nevertheless, the results of data processing showed that all of the existing service attributes have a negative gap score, and this was a weakness that exists in the implementation of Industrial Internship. The attribute that has a negative score gap and category M, then the quality of the attributes must be improved to meet the consumer expectations. While the attributes with category O and negative score, then the quality of the attributes must be improved so that it can exceed the desire of consumers.

The result of QFD Level one is the attributes rank results from servqual and Kano model analysis

Table 4. Quality Function Deployment Level

Technical Response	Weight
The appearance of Industrial Internship participants and teachers	4,2
The ease implementation Industrial practice	16,7
Participants' readiness	20,0
Attention from the school	21,3
Involved Industri Company Reputation	8,1
Industry commitment to support the Industrial Internship Program	18,1
The School Reputation	11,6

The QFD Level 1 results show that attributes with the highest importance value are attributes of attention from the school. The 2nd level HOQ matrix explains what the engineering parameters are and how to meet them. This matrix is made based on the merger of data processing from the determination of the weight of normalization to the interaction of process requirements Based on HOQ level 2 results, the priority of the process needs are to provide facilities for the participants, to determine feasibility standards for involved companies, moral development for the participants, to determine the duties and obligations of supervising teachers, to write SOP of pre-implementation, to determine the standards for Industrial Internship activities, to establish the policy of internship participants' wearpack.. A summary of the results can be seen in Table 5.

Table 5. The Relationship between *Servqua Method, Kano* Model, and QFD

Servqual	*Kano Model*	QFD
The result of the Servqual method is that all service attributes in industrial practice are still below consumer expectations. This indicated by the gap in all attributes. This means all attributes must be entered to continue being further processed with the Kano model.	The Kano model indicates that there are only two attributes that have an attractive or delighter category. Delighter serves as an attribute of consumer satisfaction that becomes an added value in the implementation of the industrial internship. There are two attributes that have indifferent categories which mean the customer is not concerned with the existence of these attributes.	Quality Function Deployment in the final form of a House of Quality has the conclusion that there are seven priorities for improving service quality in the industrial internship

4. Conclusion and Suggestions

4.1 Conclusion

The real condition of the implementation of the Industrial Internship is considered poor. Because in all the attributes expressed in the statements, there is still a gap between perception and expectation. When customer expectations are not met then customer satisfaction is not achieved. It means in this case, customer satisfaction has not been achieved although there are attributes with small gaps. There is still a gap between perception and expectation measured using servqual method and the Kano model. This was proven by negative values of customer satisfaction scores in all attributes. All attributes have an important capacity to improve the quality of the implementation of Industrial Internship especially the attributes that have the category attractive, one dimensional, and must be. All Kano attributes are continued to be combined with servqual to find the value of adjusted importance, the priorities that school management needs to do in improving the quality of the implementation of Industrial Internship are: to provide facilities and infrastructure for Industrial Internship participants, to determine feasibility standards of the companies involved in the Industrial Internship Program, moral development for the participants, to determine the duties and obligations of supervising teachers, to determine SOP of pre-implementation of the program, to determine the standard activities of Industrial Internship participants, and to do establish the policy of internship participants' wearpack.

4.2 Suggestion

To improve the service quality in Industrial Internship, the school should prioritize improvements on: providing facilities and infrastructure for Industrial Internship participants, determining the feasibility standards of the companies involved in the Industrial Internship, moral development for Industrial Internship participants, determining duties and obligations of supervising teachers, make SOP of pre-implementation of the Industrial Internship, determining the standard activities of Industrial Internship participants, establishing the policy of internship participants' wearpack. In the Kano Model, where two indifferent categories are found. It is better for the school to study the cause of the existence of the category. It is necessary to conduct further research on the application of Servqual,

Kano Model and Quality Function Deployment method in VHS 2 Klaten to improve the quality of Industrial Internship services.

5. References

[1] Blinova, T., Bylina., S & Rusanovskiy, V 2015 Vocational education in the system of determinants of reducing youth unemployment: Interregional Comparisons (*Procedia-Social and behavioral Sciences*, 214, 526-534)
[2] Surachim, A 2016 *Efektivitas pembelajaran pola pendidikan sistem Ganda* (Bandung: Alfabeta)
[3] Chuia, T.B, Ahmad M.S., Bassim., F.A., Zaimi, N.A 2015 Evaluation of service quality of private higher education using service improvement matrix (*Procedia - Social and Behavioral Sciences* 224 132 – 140)
[4] Ibrahim, M.Z., Rahman, M.N.A, & Yasin, R.M. 2012 Assesing students perceptions of service quality in technical Educational and vocational training (TEVT) institution in Malaysia (*Procedia- Social and Behavioral Science*, 56, 272-283)
[5] Sallis, Edward 2015 Total quality management in education (Yogyakarta: IRCiSoD)
[6] Fardiana, F&Suwignjo, P 2006 *Peningkatan kualitas pelayanan pada MMT-ITS dengan menggunakan integrasi servqual, kano dan QFD.* (Prosiding Seminar Nasional Manajemen Teknologi IV)
[7] Tan, K.C& Pawitra, T.A 2001 managing service quality: integrating servqual and kano's model into QFD for service excellence development,(vol 11, no.6, h.418-430)
[8] Wibisono, D 2013 *Panduan penyusunan skripsi, tesis & disertasi* (Yogyakarta: Penerbit Andi)
[9] Arikunto, S 2013 *Prosedur penelitian suatu pendekatan praktik* (Jakarta: Rineka Cipta)

IOP Conf. Series: Materials Science and Engineering **535** (2019) 012007 doi:10.1088/1757-899X/535/1/012007

The development of realia for the planetary gear unit to improve students' understanding on the basics of automatic transmission

M Wakid[1], and Tafakur[2]

[1,2]Department of Automotive Engineering Education, Yogyakarta State University, Yogyakarta, Indonesia

E-mail: wakidm@uny.ac.id

Abstract. This study aims to design and develop a real model of a planetary gear unit providing three kinds of simulation for the basic configurations of the planetary gear unit in which one component member is locked, two members are held together, and two input members works. This is a research and development (R&D) to develop a realia of the planetary gear unit for the learning of automatic transmission basics. The realia is expected to demonstrate how the configurations in which one component member is locked, two members are held together, and two input members moves. This study was carried out through ten stages, namely analyzing potentials and problems, needs analysis, product design, focus group discussion, product development, pilot study (small-scale field test), product revision, large-scale field test, product revision, and final product. The data were collected by questionnaires and were analyzed quantitatively. The product of this study is learning media in the form of realia made of acrylic so that the components are clearly visible and the work of basic configurations of the three gears in the planetary gear unit is easy to understand. These basic arangements can be operationalized on the learning media by applying brakes to the ring, carrier, and sun gears and applying cluthes between the sun and carrier gears, the sun and ring gears, and the carrier and ring gears. Evaluated from its performance, the realia of the planetary gear unit can work well. The shaft can rotate well though some friction generated, the clutch and brake can work on its function, as well as the constituent gears can rotate as its real version does. This is reinforced by the responses from the material and media experts that reveal the product meets the very feasible category.

1. Introduction

It has been understood that various factors affect the quality of automotive learning, such as those from the student, teacher, facility, environment, and institution to name a few. Regarding this, active and creative students that are sufficiently assisted will produce effective learning outputs. However, some learning contraints still exist, one of which is on the learning of chasis, particularly on the Automotive Power Transfer System. Based on the observation on the automatic transmission materials delivered by teachers in some Vocational High Schools (VHSs) in Yogyakarta Special Region, lectures in the Automotive Engineering Education Department (AEED) of Faculty of Engineering (FE), Yogyakarta State University (YSU), the professional trainers at Yogyakarta Technical and Vocational Education and Training (TVET) Center, teachers in the Teacher Professional Education and Training Program (*PLPG*) in the field of ligh vehicle, and teachers in the Teacher Professional Program (*PPG*) in the same field, it was proved that the leacturers' and teachers' understanding on the concept of the planetary gear unit is still low because the material is challenging.

The problem that arises is possibly caused by the difficult to imagine how a planetary gear unit works when the explanation is unaccompanied by any animation videos. Many students also required some real learning media that can be simulated directly during the learning process. In the AEED and some VHSs formerly assisted by the World Bank programs, real media for the planetary gear had been developed. However, they are only able to show one configuration model that holds one main component of the planetary gear with two main components configured for holding the ring and carrier gear.

Another problem is the difficulty of studying the automatic transmission system for both hydraulic and electro hydraulic systems. Suyitno states that real media is needed to show how the systems work to help student understanding [1]. Whereas if to overcome the problem of limited space and time, digital media in the form e-leanirng can be used [2]. As there has been no real media for learning the planetary gear sold in markets or e-commerce, new solutions should be found to ensure the teaching and learning process reaches its goals.

Based on the discussion above, therefore the problems are formulated as follows. (1) How is the realia capable of providing simulation variations on the basic configurations of the planetary gear unit? (2) How is the feasibility of the realia of planetary gear unit in terms of its functions? Based on the formulation of these problems, the objectives of this study are: (1) to design and develop a real model of planetary gear unit that can provide three kinds of simulations of the basic configurations of the planetary gear unit including the configuration of one locked member, two members held together, and two input members moving. (2) to describe the feasibility of the realia of planetary gear unit as the learning media.

Kustandi and Suctipto describe learning media as instruments that support the teaching and learning process and serve to make the information delivered more understandable [3]. Hamalik in Arsyad adds that the use of these media can encourage students' motivation and interest in something new, improve their motivation and involvement in learning, and even bring about phychological stimuli to them. With regards to this, Sudjana and Rivai add that learning media function to (1) encourage learning motivation, (2) improve the clarity of learning material delivery, (3) vary the learning methods and avoid students' boredom, (4) improve students' active participation by observing and performing instead of listening to the explanation, and (5) transfer some abstract theories into factual ones and the complex theories into simple ones [4]. Similar opinion was expressed by Susilana and Riyana that the see the benefits of learning media as: (1) making concrete some abstract concepts, (2) presenting objects that are too dangerous or difficult to obtain, (3) showing objects that are too large or small, and (4) showing movements that are too fast or slow [5].

In relation to this, many alternative forms of media have been used in learning, one of which is realia or objects from the real life. Anitah describes that realia resemble an actual object in their intact form. Realia is a real-like object that is used as a learning resource [6]. Sudjana also adds that the use of real objects in the teaching and learning process mainly aims to introduce a particular unit of learning, the working process of a particular object of study and other aspects required [7].

Realia or real life objects are visual aids in learning, namely real models or real objects. Sumantri and Permana suggest that in general concrete media function as (a) instruments to actualize effective teaching and learning situation, (b) integral parts of the overall teaching situation, (c) proofs of actual basics and abstract concepts so that can make verbal explanations easier to understand, (d) instruments to improve students' learning motivation, (e) enhancing tools for the quality of teaching and learning[8].

Ibrahim and Syaodih state that real object media will provide stimuli that are very important for students in learning various things, especially those concerning the development of certain skills [9]. Additionally, Subari states that teaching aids are used by teachers to demonstrate teaching materials in order to provide a clear understanding of the lesson [10]. Therefore, with these properties or realia, the automotive chassis learning on the automatic transmission section will be greatly supported.

2. Research Method

This is a Research & Development (R&D) to develop and validate the realia planetary gear unit as learning media with the adjustment and size suitable with the learning needs. This study was carried out through 10 stages, namely: analyzing potentials and problems, needs analysis, product design, focus group discussions, product development, pilot study (small-scale field test), product revision, large-scale field test, product revision, and final product presentation. These stages were adapted from the steps of Sugiyono [11].

The subjects of this study were students who had attended the course on the Power Transfer System. This research was carried out in the Automotive Engineering Education Department (AEED) of FE YSU. The data were collected by observation sheets and questionnaires. The questinonnaires were used to gather the data form the material experts and from students. Then, the data were analyzed by means of quantitative descriptive analysis techniques. The scores obtained were consecutively compared to the table of the eligibility criteria as presented in Table 1.

Table 1. The Product Feasibility Criteria

No	Score	Category
1	1.00 – 2.50	Completely Unfeasible
2	2.51 – 5.00	Somewhat Feasible
3	5.01 – 6.50	Feasible
4	6.51 – 8.00	Completely Feasible

3. Results and Discussion

Based on the research objectives, this research was carried out to (1) design and develop a real model of the planetary gear unit which provides three kinds of simulation for the basic configurations of the planetary gear unit, that is the configuration of one locked member, two members held together, and two input members working. It also aimed to describe the performance of the real planetary gear unit model in terms of its feasibility as learning media. The stages of development that have been carried out include the followings.

3.1. Needs Analysis

As in the stage of identification of potentials and problems the problem of learning automatic transmission in the planetary gear unit has been now, the next is planning the development. The developed media needed should be able to (1) display the planetary gear unit components, (2) simulate the rotation of each planetary gear unit part, (3) show the clutch and brake configuration on the planetary gear unit, and (4) be easily used.

3.2. Product Design

The development of the realia design was carried out with a computer program to determine the construction, dimension, and work simulation of the developed media.

Penelitian Dosen Muda Fakultas Teknik UNY Tahun 2017

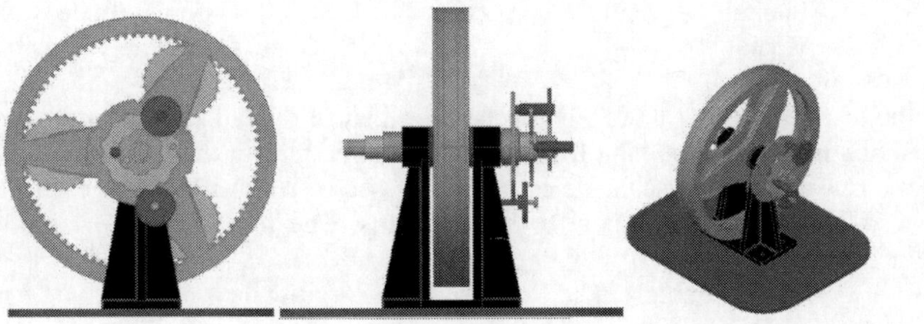

Figure 1. Product Design

In addition, the product materials were also determined based on the needs. To be able to show the performance of the components clearly, the main material chosen for the ring gears was acrylic because of its transparent nature. For gears, shafts, brake, and clutch mechanisms, the best material to be used was metal. The design was then discussed by the lecturer team to validate the product design that has been determined.

3.3. Product Development
In this process, the product development was carried out in accordance with the predetermined design. There were several plan changes, including the gear materials that were all made of acrylic because of an obstacle in the manufacture of gears from metal. Considering the limited time for production, it was decided that the gears were made of acrylic.

3.4. Product Evaluation
After the product was developed, it was evaluated by the media experts and material experts. The results of product validations are presented in Table 2.

Table 2. The Product Validation Results

Item	Media Experts	Material Experts
Score	109	105
Maximum Score	120	112
Number of Item	15	14
Means	7.27	7.5
Catedory	Complety Feasible	Complety Feasible
Suggestions	1. A bearing should be installed to prevent sliding friction. 2. The bushing can be made of metal to maintain the product durability.	1. A rotation sign for the input/output shaft should be given. 2. The components should be lubricated.

Based on the results of validation by material experts and media experts, the developed media was found to meet the criteria of feseable learning media and suit the planetary gear unit material. However, suggestions were given, namely: (1) giving lubricants to moving components, and (2) giving a description on the input and output on the realia. After the revision, the realia was then tested in small groups of students (the small-scale tryout).

3.5. The Field Tests
The small-scale tryout involved ten practical course students as respondents. The test results of realia media are presented in Table 3.

Table 3. The Results of the Small-scale Field Test

Aspect	Total Score	Max Score	Means	Category
Material	509	8	6.36	Completely Feasible
Media	445	8	6.36	Completely Feasible

Based on the test results in Table 3, it can be seen that in overall the realia developed is completely feasibility. Each indicator meets the feasibility criteria, both from the media and material aspects. The next step taken was the large-scale field test by the use of realia in a large class. The respondents of this field test were twenty practical course students. The results for the large-scale field test are presented in Table 4.

Table 4. The Results of the Large-scale Field Test

Aspect	Total Score	Max Score	Means	Category
Material	1,106	8	6.91	Completely Feasible
Media	964	8	6.94	Completely Feasible

Based on the results of the large-scale field test, it can be seen that the overall score obtained is high with a means of 6.94, thus the media developed is categorized as completely feasible. Therefore, the learning media can be used to support the teaching and learning process. This is in line with the statement that the use of media in learning is very helpful in improving learning outcomes, one of which was revealed by Hertanto in the results of his research [12].

3.6. Product Revisions

After the product field tests, the realia was revised to provide a sign on the rotating objects so that it is easier for the students to observe the rotation. In practice, this planetary gear unit is expected to facilitate students' understanding of the planetary gear unit configurations. As stated earlier, the work of automatic transmission on vehicles utilizes the planetary gear unit principle. A realia that is easy to use and attractive, and functions as the original object threfore will help the students in understanding the basic principles of the automatic transmission. With this media, students are expected to be more motivated to learn the automatic transmission materials because some students were found to have low understandin on the materials.

As revised, the realia components are clearly visible so that the media can help students to identify planetary gear components, the work of the clutch and brake setup, and the gear ratio on the planetary gears. Thus, if the abstract planetary gear unit that can be delivered in terms of its working proces, animation, and image will be more undestandable when such realia is used. This media can help learning as what the original object does so that the students can easily understand the work of the planetary gear unit through the model. Consequently, when the students understand the planetary gear concept, they will be easier to understand the automatic transmission. This is in accordance with the statement that to achieve effective learning in vocational education, competency must be achieved well before going to the next competency [13].

4. Conclusion

Based on the results of data analysis and the discussion on the development of the realia of the planetary gear unit, some conclusions presented are as follows. (1) This realia is made of acrylic material so that the components are clearly visible and the work of the gears can be easily understood. The realia presents the three rounds of the planetary gear basic arangements by applying a brake to the ring, carrier, and sun gear and applying a cluth between sun and carrier gears, between the sun and ring gears, and between the carrier and ring gear. (2) Evaluated from its performance, it is known that the realia of the planetary gear unit apparently works well. The shaft rotates well though it producess friction, the clutch and brake work on its function, as well as the rotating gears move as on its real version. This is reinforced by the response from material and media experts, and from the users who state that the realia can work well, and that it meets the Completely Feasible category.

5. Refereces

[1] Suyitno (2018). Developmnet of learning Media for the course of-Two-Stroke Gasoline Motors to Improve Students Learning Outcomes. *JPTK*, 24 (1), May 2018, pp. 83-90.
[2] Hidayat, Nur. (2018), Developing E-learning Media with the Contiguity Principle for the Subject of AutoCad. *JPTK,* 24 (1), May 2018, pp. 72-82.
[3] Kustandi C. and Suctipto B. (2011). *Media Pembelajaran Manual dan Digital.* Bogor: Ghalia Indonesia.
[4] Sudjana, N. and Rivai, A. (2010). *Media Pengajaran*. Bandung: Sinar Baru Algensindo.
[5] Susilana, R. and Riyana, C. (2009). *Media Pembelajaran*. Bandung: CV Wacana Prima.
[6] Anitah, S. (2009). *Teknologi Pembelajaran*. Surakarta: Universitas Sebelas Maret.
[7] Sudjana, N.. (2007). *Media Pengajaran.* Bandung: Sinar Baru Algesindo.
[8] Sumantri, M. and Permana, J. (2001). *Strategi Belajar Mengajar*. Bandung: CV Maulana.
[9] Ibrahim and Syaodih. (2003). *Perencanaan Pengajaran*. Jakarta: Rineka Cipta.
[10] Subari. (1994). *Supervisi Pendidikan*. Jakarta: Bumi Aksara.
[11] Sugiyono. (2011). *Metode Penelitian Pendidikan (Pendekatan Kuantitatif, Kualitatif, dan R&D).* Bandung: Alfabeta.
[12] Hertanto, D.B.. (2011) *Upaya Peningkatan Kualitas Kuliah Jaringan Komputer melalui Penenarapan Media Pembelajaran Packet Tracer 5.0.* Retrieved from https://journal.uny.ac.id/index.php/jptk/article/view/7753/6670.
[13] Blank, Wiliam E. (1982). *Handbook for Developing Competency-based Training Programs.* Englewood CliffNew Jersey: Prentice Hall.

IOP Conf. Series: Materials Science and Engineering **535** (2019) 012008 doi:10.1088/1757-899X/535/1/012008

The effect of NaOH concentration variation in the process of paper making from bamboo fiber

Dulmalik[1], Dwiky MI[2],N Kumala S[3], Fety IR[4], FN Khaliq[5]

[1,2,3,5]Department of Chemical Engineering, Faculty of Industrial Technology, Islamic University of Indonesia, Yogyakarta, Indonesia
[4]Department of Industrial Engineering, Faculty of Industrial Technology Islamic University of Indonesia, Yogyakarta, Indonesia

E-mail: 815210102@uii.ac.id

Abstract. Bamboo cropping period is quite short compared to wood, but fiber and bamboo pulp have not been optimally used as a substitute for wood in composite industry, such as fiber board. It is worth to be investigated because of the availability of wood is declining. Therefore, this research aimed at revealing the use of fiber and bamboo pulp (G. apus) for fiberboard. The study was conducted using bamboo fibers from bamboo strips cooked with soda process and bamboo pulp from chips of bamboo cooked with soda process and Kraft process. Composite of bamboo fiber is made with a variety of fiber weight fraction to the epoxy resin matrix. Furthermore composite of bamboo pulp is made under optimal condition for making fiber composite. Based on the test results, it is known that the quality of the bamboo fiber composite from bamboo strips cooked with soda process is better than the composite of bamboo pulp. Besides, the produced bamboo fiber composites can be categorized as the high-density fiberboard meanwhile for the aspect of water content, water absorption, length and thickness changes, flexural strength and modulus of elasticity are in accordance with applicable standard for fiberboard (SNI 01-4449 - 2006).

1. Introduction

In line with the advancement of technology, the demand of paper is currently increasing for schools, offices, companies, households and others. Paper is a thin material made of wood and it has even become a daily necessity as a medium for writing, drawing, printing, wrapping and also used as a cleaning tool. The increasing request of paper in everyday life effects on massive deforestation for the raw material of papermaking. It impacts on the environment, as the lack of water consumption on the ground that can cause flooding, landslides and even declining oxygen contain in the air which can cause global warming. To overcome this problem, it is crucial to find an alternative solution by changing he raw material of paper from pine tree fiber into bamboo tree, especially Bamboo *tali* (rope) or called bamboo apus. It is a type of bamboo commonly used for making paper because the fibers are soft, strong and flexible so they are not easily broken. Usually, the process of making pulp is through the process of craft, sulfite, or soda. Considering the potential of bamboo which can be made into pulp and the lack of research on bamboo-based pulps with the Soda process, it is beneficial to investigate the use of bamboo as a raw material for pulp making.

2. Literature Review

Bamboo apus (Gigantochloa apus) is a type of bamboo that is widespread in Indonesia and tropical Asia. This bamboo is widely cultivated for the manufacture of handicraft raw materials. The taxonomic system for bamboo rope or bamboo apus is Kingdom: Plantae , Divisi: Angiospermae, Klas: Liliopsida,Ordo : Poales, Famili : Poaceae, Genus: Gigantochloa, Spesies : Gigantochloa apus (Bl. ex (Schult F. Kurz.)

The origin of bamboo apus is from Burma and it has spreaded widely throughout the Indonesian archipelago. This kind of Bamboo grows well in the tropics in the lowlands to the mountains with altitudes up to 1,000 m above sea level. The of bamboo apus is planted with the rhizomes roots and it can also be propagated by cutting the reed. Berlian and Rahayu state that bamboo apus can reach up to 20 meters in height [1,2,3]. The color of bamboo stem is green to yellowish. It has no branch at the bottom part. The stem diameter is between 2.5 to 15 cm, the wall thickness is 3 to 15 mm, and the length of the segment or reed is 45 to 65 cm. Utilization of bamboo apus culms is between 3 and 15 meters of which flexural strength is 502.3 - 1240.3 kg / cm2, modulus of flexural elasticity is 57,515 - 121,334 kg / cm2, flexural strength of 1,231 - 2,859 kg / cm2, and firmness press of 505,3 - 521,3 kg / cm2. The mechanical properties of bamboo apus without books are bigger than those with books [4,5].

2.1. Chemical Process

The chemical pulping process is a pulping process that uses chemicals as the main ingredient to dissolve unwanted parts of bamboo. The principle of the chemical pulping process is to degrade and dissolve lignin so that the fibers contained in the raw material are easily removed. There are three types of chemical pulping processes, namely soda process, process of sulfate or kraft, and sulfite process, each process uses a different cooking solution. The sulfate process and the soda process are both called alkaline processes while the sulfite process called as acid process. Pulping process as a basic process, where the cooked solution used for soda process is NaOH, while the sulfate process used cooking solution NaOH, Na2S, and Na2C03. and for the sulfite (acid) process a sulfite salt cooking solution is used. The criteria for the success of the chemical pulping process are product quality and high pulp recovery with little water use as well as high chemical recovery rates. In practice, the methods of chemical pulping have succeeded in separating most of the lignin but also dissolving certain amounts of hemicellulose and cellulose to make pulp recovery is relatively low compared to mechanical pulping. The acquisition of chemical pulp is usually in the range of 40-50%.

2.2. Sodium Hydroxide (HaOH)

NaOH is a strong alkaline chemical. It is also known as Caustic soda in the form of solids (white chips / crystals) for trading. Sometimes, it called as soda, natronloog, white kostik, or sodium hydrate. NaOH widely used in making rayon, paper, soap, detergent, textile processing, etc. (Guidelines for Textile Chemistry Practicum: 8). In the process of cooking cellulose natural fiber, the function of NaOH is to dissolve fat and dirt contained in the fiber so that the fiber becomes clean (Soeparman, 1967: 265). However, because NaOH is corrosive, it may damages materials such as textiles, leather, or paper. So, the use of it must consider its concentration (Utari B, 1986: 12). In this study, NaOH concentration was 5%, 10% and 15%.

3. Research Method

3.1. Data Collection Technique
This research was using literature study method by collecting data by reading from the internet sources and books related to the problem being studied as well as laboratory study methods. It also used direct experimental data on the studied sample and test samples of experimental results in the laboratory.

3.2. Sampling metodology
The sample division conducted in this study was using: (1) 50: 10 gr bamboo fiber samples and used paper, 2000 mL of aquadest and NaOH of 5%. (2) 50: 10 gr sample of rami fiber and used paper, 2000 mL of aquadest and NaOH of 10%. (3) 50: 10gr sample of rami fiber and used paper, 2000 mL of aquadest and NaOH 15%

3.3. Research Framework

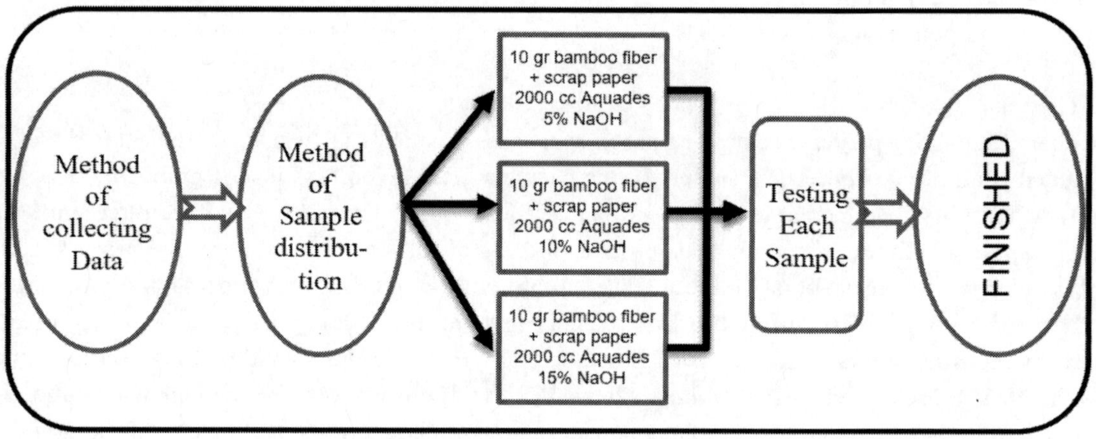

Figure 1. Research Framework

4. Result and Discussion

4.1. Advantages and Benefits
There are several advantages and benefits of bamboo fiber rope for paper making, including: (1) Flexibility - high elasticity so that it is not easily damaged, (2) Resilient-elastic so that the formation of paper becomes stronger (3) The fiber is long so it has great potential to be formed with various variations (4) It is environmentally friendly so that it easily breaks down in the soil.

4.2. Application
In the process of kraft paper making, bamboo apus fiber is used as the main raw material. This kind of bamboo has high flexibility or elasticity which is not easily broken and resilient so that it is very appropriate to be used for paper making.

This study is using chemical process (soda process), i.e pulping process with chemicals as the main ingredient to dissolve discarded parts of bamboo. The principle of this chemical pulping process is to degrade and to dissolve lignin so that the fibers contained in the raw material can be easily removed. The pulping process with basic processes of which the pulp is cooked using NaOH solution. The criteria

IOP Conf. Series: Materials Science and Engineering **535** (2019) 012008 doi:10.1088/1757-899X/535/1/012008

of the successful chemical pulping process are on product quality and high pulp recovery with little water use as well as high chemical recovery rates. Based on research on paper which consists of a pulping process that uses a mixture of bamboo fiber and used paper with a variable of (50: 10%), and variation of bamboo fiber length ± 3 cm. The raw material for making paper is cellulose which is chemically treated, rinsed, decomposed, bleached formed into sheets after pressing and drying. The optimum cooking time in the delignification process was about 60-120 minutes with the lignin content fixed after this range. The longer the cooking time, the lignin content in the pulp is high, because the lignin that has been separated from the raw pulp due to reduced concentration of NaOH will re-join with the fused raw pulp and it is difficult to separate. The immersion time is varied for the paper formation in order to softening the used paper to form into pulp.

Table 1. Test result of the paper with NaOH concentration of 5%,10%, and 15%.

Sample	Test of	Test score of flexural strength and modulus of elasticity		Thickness Test Score (mm)
		Flexural Strength Score (Gr)	Modulus of Elasticity Score (%)	
5 ml	1	900.00	0.60	1,222
	2	900.00	0.466	1,284
	3	800.00	0.466	1,290
Mean score		866.667	0.511	1,265
10 ml	1	3100.00	0.866	2,232
	2	3200.00	0.733	2,546
	3	3100.00	0.666	2,345
Mean score		3133.333	0.755	2,374
15 ml	1	2700. 00	0.666	1,109
	2	2600. 00	0.60	1,130
	3	2200.00	0.60	1,193
Mean score		2500. 00	0.22	1,144

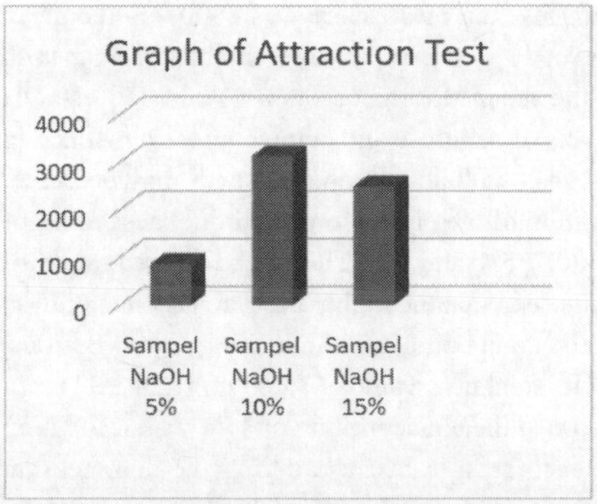

Figure 2. The relationship between the percentage of NaOH concentration and the flexural strength of the paper

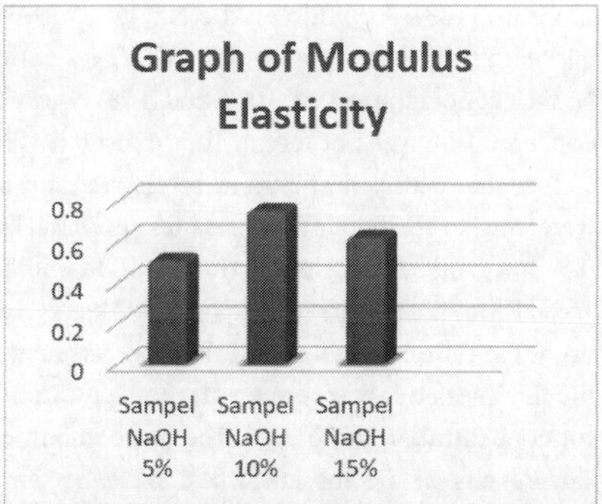

Figure 3. The relationship between the percentage of NaOH concentration and modulus of elasticity.

Figures of Paper surface

Figure 4. Sample NAOH 5 %

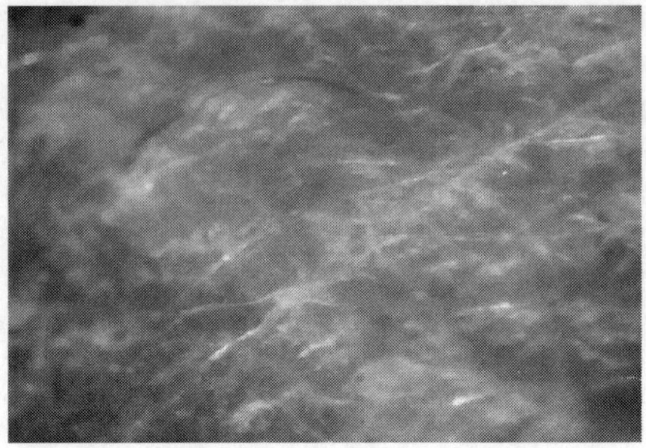

Figure 5. Sample NAOH 10%

Figure 6. Sample NAOH 15%

Tenso Lab is used to test the flexural strength and paper modulus of elasticity of the paper. It works by clipping a sample of 2x20 cm paper into a clamp that is already available in the tool. Based on the test results of the lab that flexural strength was conducted test three times, the average paper attractiveness at 5% NaOH concentration was 866,667, the 10% NaOH concentration obtained the mean score of 3133,333 while on the NaOH concentration of 15%, the mean score was 2500,000 the mean gap with NaOH concentration of 5% and a concentration of 10% of 2266.666. Meanwhile, the gap with NaOH concentration of 10% and 15% was 633,333, the mean difference on 10% and 15% NaOH concentration was not too far. It is caused by the less levels of cellulose and hemicellulose contained in a fiber that cause the fibers to be shorter and easier to break so that the paper strength may decrease. Referring to the paper thickness test in NaOH concentration of 5%, it was obtained thickness score of 0.511% while for concentration of 10%, the score was 0.755% and 0.622% for the concentration of 15%. From the results of the paper thickness test, it can be seen that the higher the NaOH concentration used, the more dissolved cellulose because cellulose is the main component in making paper. Besides, the thickness test can be seen at a concentration of 5% with thickness value of 1.265 mm and 2.374 mm for concentration of 10%. Furthermore, the thickness score in the concentration of 15% was 1.144 mm. The results of the thickness test were not similar for each sample because during the compression process the researcher used a simple paper printing device made by the researcher which was done manually. The lab graph test showed the score of flexural strength was not linear following the Langmuir equation so that it can be concluded that the optimum NaOH concentration in paper making is at the addition of 10% NaOH concentration. In this case, with NaOH concentration of 10%, it was obtained the highest flexural strength.

5. Conclusion

(1) The bamboo stem mainly contain cellulose. It is a polysaccharide composed of D-glucose monomers. It has three hydroxyl groups that can be substituted because the basic material for paper making is cellulose. It means bamboo can be used as material for paper and cellulose is a photosynthetic product of plants. (2) Bamboo apus has special features on its stems and leaves in which the fiber is smooth, long and strong so that it is suitable as the raw material of the paper making. (2)The longer the cooking time, the bamboo fiber can be shorter and fragile so that the flexural strength the paper decline, the long cooking time may cause the cellulose to break down so that the produced yield and pulp is low. (3) Based on the graph of lab test, the score of flexural strength and modulus of elasticity is not linear with Langmuir equation. It can be concluded that the optimum NaOH concentration in making paper is in the addition of NaOH concentration of 10%. (4) In 5% of NaOH concentration and 10 g of used paper mixture, the flexural strength test score was 866.667 g with the flexural test score was 0.511% and the thickness test was 1.265 mm. (5) In 10% of NaOH concentration and 10 g of used paper mixture, the flexural strength test score was 3133,333 g with the flexural test score was 0,755% and the thickness test was 2,374 mm. (6) In 15% of NaOH concentration and 10 g of used paper mixture, the flexural strength test score was 2500,000 g with the flexural test score was 0,622% and the thickness test was 1.144 mm.

For further research, it important to add a sample of NaOH concentration (>3 samples) with different concentration in the range of adjacent distances to obtain more accurate results.

6. References

[1] Berlian, Nur V. A. dan Estu Rahayu. 1995. Jenis dan Prospek Bisnis Bambu. Penebar Swadaya. Jakarta.

[2] Dransfield, S and E. A Widjaya. 1995. Plant Reourch of South East Asia (PROSEA) no. 7: Bamboo, Bachuys Publisher. Leiden

[3] Liese, W. 1985. Anatomy of Bamboo Proceding Workshop Bamboo Research in Asia, Singapore 28-30 May 1980. International Development Research Center. Ottawa.

[4] Heyne. 1987. Tumbuhan Berguna Indonesia. Badan penelitian dan Pengembangan Kehutanan. Bogor.

[5] Sastrapraja, S, E A. Widjaja, S. Prawiroatmodjo, ddan S. Soenarko. 1980. Beberapa Jenis Bambu. Lembaga Biologi Nasional-LIPI. Bogor.

IOP Conf. Series: Materials Science and Engineering **535** (2019) 012009 doi:10.1088/1757-899X/535/1/012009

Android-based applications on teaching skills based on TPACK analysis

P Utami[1], FR Pahlevi[2], D Santoso[3], N Fajaryati[4], B Destiana[5] and ME Ismail[6]

[1,2,3,4,5]Universitas Negeri Yogyakarta, Karangmalang, Sleman, Yogyakarta, Indonesia
[6]University Tun Hussein Onn Malaysia, 86400 Parit Raja, Johor, Malaysia

E-mail: pipitutami@uny.ac.id

Abstract. Teaching skills learned in micro-teaching are important competencies in the learning management. Dependence of other parties, both lecturers and colleagues teaching during the assessment process of teaching practice and lack of media that integrates the concept of material, examples, and practices are obstacles in mastering teaching skills. This article describes the specifications needed of an Android-based learning tool (media) for microteaching ("Microteaching") courses and analyzes the performance of these learning media. The development phase begins with Analysis, Design, Development, and Evaluation. The results of the analysis referring to TPACK and the design showed that the specifications of the Android-based media consisted of two main menus namely material (containing various multimedia elements) and assessment (besides text-based assessment also contained speech recognition elements). The performance of "Microteaching" in functionality is 100% successful, while the feasibility test at the evaluation stage not only for media experts, but material experts and users indicates that the media is suitable for use. Furthermore, results of the assessment show that the media is able to depend on the assessment, except from the aspect of verbal aspects. Integration of learning concept material through mind maps, technical examples related to verbal teaching skills and videos to provide concrete description of how to teach (both verbal and non-verbal).

1. Introduction

In the education study program, micro teaching is a course that must be followed by students to get information and experience in ways to teach technically in the classroom. Micro-teaching is an early stage training in the formation of teaching competencies for prospective teacher [1]. Micro-Teaching is a course that has a very important role to equip prospective teacher with teaching skills[2]. To learn teaching skills, the teacher training techniques is microteaching[3]. Especially, it is a technique to train the prospective teachers[4]. Micro-teaching is a powerful way in teacher training to improve the instructional experience of prospective teachers and has been adopted in a number of teacher education institutions that are committed to producing qualified future teacher candidates[5]. In a limited time, especially students of technical education study program get teaching experience in front of the class to apply pedagogical theory and engineering practice. Next, through the teaching practice program at school, students can implement the teaching skills learned in micro-teaching courses. Micro-teaching is an efficient technique that has a positive impact on the competence of teaching students when the student practices teaching in a real class (school)[5]. Mastery of teaching skills is important to be achieved before the implementation of teaching practice in schools. Therefore, micro-teaching learning plays an important role in preparing future teachers, especially in preparing mastery of teaching skills.

One of the implications of the successful achievement of micro-teaching learning outcomes in education study programs is through successful teaching practices in schools. In the period of teaching practice in schools from 2015 to 2017 at one of Vocational High School in Yogyakarta, some aspects of teaching skills owned by students are still lacking or even not appearing and there are indications that students are not confident when teaching in class[6]. The interviews showed several things, namely: (1) aspects of teaching skills that were rated at least were classroom management skills; (2) skills in conducting learning variations are also considered lacking which is indicated by the frequent giving of written assignments to students (there is no variation in student activity); (3) skills to teach small groups and individuals are still lacking which is indicated by the lack of clarity in the delivery of the objectives of group learning activities, the lack of a positive response and often the students sit in the teacher's seat (not approaching and guiding); (4) do not pay attention to aspects of teaching skills on lesson plans; and (5) self-distrust indicated by the implementation of teaching practices that always invite other students in the class[6]. Especially the findings of the fact that there are students who are not confident are contrary to the results of research that show that micro-teaching is able to increase self-awareness and confidence, and also supports the discovery of teaching styles for students[7]. The findings of students' unpreparedness in the practice of teaching in school indicate problems in micro-teaching learning.

Based on the lesson plan at Micro Learning in the Faculty of Engineering, Yogyakarta State University shows that from 16 meetings, ten teaching practices were held, namely one limited teaching skills practice meeting, three meetings of integrated skills in theoretical subjects, three meetings of integrated skills practice subjects in the laboratory, and three meetings on the practice of integrated skills in practical subjects in the workshop[8]. Interviews with lecturers showed the conclusion that the assessment of one aspect of students' teaching skills in one class took 2 meetings and students could practice teaching class 6 times. In addition, from the 6 times the practice, students did not fully practice the overall teaching skills and some teaching skills were not trained repeatedly or only one practice. Teaching skills require habituation, so the number of repetition exercises can be a factor that determines the flexibility of teaching.

Other findings on micro-teaching learning in the classroom are media on teaching skills material in the form of mastery of concepts, examples and direct practices for students that are integrated and easily accessible for students not yet realized. The deepening of material and teaching skills practices for students is very dependent on face-to-face learning in class. One of the obstacles in micro-learning is the dependence of the role of lecturers as assessors of student teaching skills. Therefore, it is necessary to make efforts to reduce the dependence of the assessment through the use of technology in learning. In 21st century learning, educators must implement technology and media to help students master 21st-century skills[9]. In addition, pre-service teachers need significant practice working with various technological tools as part of their teacher- education program[10]. One technology that can be utilized is a mobile application, a software system that functions on a mobile device[11]. When used in learning, it can be termed a mobile learning application that is a form of learning on mobile telecommunications devices with mobile networks. Benefits that can be taken from the use of mobile learning applications include: (1) rapid communication; (2) portable and easy to use; (3) provide opportunities to access learning at any time and anywhere; (4) learning material can be remembered more by users because it is learned according to the user's learning desires; (5) facilitate user interactivity; (6) costs that tend to be cheaper because there are generally users already have a mobile device[12]; (7) having more privacy[13]; and (8) more flexible assessment[14]. One device that supports mobile learning is a smartphone.

Smartphone users in 2014 are 1.57 billion and continue to increase every year until predicted in 2020, smartphone users totaling 2.87 billion[15]. Students are one component of smartphone users. Current students include generation Z, the generation born after the birth of the internet. In Indonesia, generation Z is a generation born in 1995 and thereafter[16]. Generation Z has better access to Information and Communication Technology (ICT) than the previous generation[17]. This fact is the basis for the importance of developing learning media that can be used on smartphones. Observation in class shows that all students have a smartphone, so the development of media on smartphones is considered easily

IOP Conf. Series: Materials Science and Engineering **535** (2019) 012009 doi:10.1088/1757-899X/535/1/012009

accessible to students. There are many operating systems that control smartphones. Android is an operating system that gets the highest sales value compared to other operating systems [18]. Some of the things that cause Android to be selected include speed factors, productivity aspects, user-friendly design creativity, flexibility and reliability [19] and because Android is a complete, open source and unpaid platform[20]. This is the basis of consideration for the development of developing learning media on smartphones using the Android operating system. One developer in developing an Android application is Android Studio. The selection of the developer is based on the official Android programming IDE from Google and has features that are easily developed by developers.

The application of the Android application as a learning media must fulfill the elements of learning media. The elements of the learning media in question include: (1) aiming to convey messages; (2) the achievement of learning objectives; (3) there is a match between the characteristics of the media and the characteristics of the subject matter presented; (4) adjusted to the form of learning activities[21]; (5) clarifying information (understanding concepts); (6) facilitate and improve learning processes and outcomes; (7) raises learning motivation; (8) direct interactivity by users according to their abilities and interests; (9) overcoming the limitations of the senses, space and time; (10) giving students the same experience[22]; (11) meet the elements of suitability or relevance, ease, attractiveness and usefulness[23]. The usability element is the most important thing in the development of mobile learning application[24] because supporting the application becomes easy to use (easy to use), increases productivity, reduces training costs, and increases user satisfaction (5), and can ensure that interaction design and Interface design is well done to support interaction (8). Usability instrument[25] was chosen because the instrument was prepared for mobile applications that have similarities with the applications developed in this study.

In general, the stages of software development consist of analysis, design, development, and evaluation. Pressman[26]stated that the first stage of software development needs to do Software Requirements Analysis. Regarding the use of technology in learning, the analysis is carried out through Technological Pedagogical Content and Knowledge (TPACK). TPACK is a technology integration framework that identifies three types of needs that need to be combined in teaching namely technology, pedagogy and knowledge material[27]. At TPACK, technology must not master learning, but technology needs to be combined with pedagogy and knowledge material. Software Requirements Analysis is done to get application specifications. Up to this stage, the analysis phase is completed and continued with the design stages. The focus of activities carried out at the design stage is data design, architectural design, interface design, and algorithm design. The development stage is a design implementation that is coding and testing errors in the program code that is created. The last stage is an evaluation. The evaluation in question is to review the application which consists of: (1) control structure testing using testing on android developers; (2) black box testing to perform functionality tests; (3) validation of media experts and material experts to conduct a feasibility test to fulfill the elements of learning media; (4) user testing to lecturers and students to conduct usability tests; and (5) product revision (application). Data retrieval uses a questionnaire with five alternative answer choices regarding the media developed. Score conversion is divided into five categories[28], which are very feasible with a score interval $X > Mi + 1.8\ Sbi$ ($X > 4.2$), feasible with Mi score intervals $+ 0.6\ Sbi < X \leqslant Mi + 1.8\ Sbi$ ($3 , 4 < X \leqslant 4.2$), enough with the Mi score interval $- 0.6\ Sbi < X \leqslant Mi + 0.6\ Sbi$ ($2.6 < X$ 4 3.4), less than the Mi-1 score interval, 8 $Sbi < X \leqslant Mi - 0.6\ Sbi$ ($1.8 < X$ 6 2.6), and not feasible with a score interval $X \leqslant Mi - 1.8\ Sbi$ ($X \leqslant 1.8$).

2. Android-based applications on basic teaching skills based on TPACK analysis

The analysis starts from defining the phenomena that occur, namely the unpreparedness of students of education study programs while practicing teaching at the school. Signs that arise are the lack of mastery of teaching skills. On the other hand, teaching skills are learned by students in micro-teaching courses. This indicates a problem in micro-teaching learning. The lack of teaching practice training and the lack of media that integrate material (concepts, examples) and assessment are the causes of problems in micro-learning, especially in the Faculty of Engineering, UNY. Teaching aid development in micro-teaching learning is needed to overcome this. The novelty to be conveyed in this paper is the use of TPACK as a method of analyzing media development. The use of technology in micro-teaching requires TPACK analysis first[29]. The TPACK analysis was carried out to obtain teaching aid in accordance with the needs of micro-teaching learning. Table 1 shows the results of the TPACK analysis performed.

Table 1. The results of the TPACK analysis

Analysis	Result
Technological Knowledge (TK)	The application of mobile learning applications on Android studio-based smartphones contains learning multimedia (text, audio, visual, involving more than two senses)
Pedagogical knowledge (PK)	a. learning tutorial for mastering the concept of material (text, mind map, image audio, video) b. assessment of concept understanding c. repetitive teaching practice (habituation).
Content knowledge (CK)	Concepts and examples of teaching skills according to the lesson plan
Pedagogical content knowledge (PCK)	a. Giving tutorial material (text, mind map, audio, video) about concepts and examples of teaching skills b. Assessment of concepts and examples of teaching skills c. Practice teaching skills
Technological content knowledge (TCK)	The use of Android-based applications with material features (text, image, audio, video), chatbot, questions, practice teaching skills and notification of training progress.
Technological pedagogical knowledge (TPK)	The use of Android-based applications with material features (text, image, audio, video), chatbot, questions, practice teaching skills and notification of progress of training through the application of hybrid models (tutorials and drills), multiple choice tests understanding concepts and exercises and assessments practice teaching skills repeatedly
TPACK	Android-based applications are used in learning with material features (text, image, audio, video), chatbot, questions, practice teaching skills and training progress notification through the application of hybrid models (tutorials and drills) for understanding the concept of microteaching material assessed through tests multiple choice and repeated practice and assessment of teaching skills

From the TPACK analysis, it is known that to overcome problems in the form of a lack of a number of repetition of teaching skills by developing teaching practice assessment features that can be done by users anytime and anywhere. The concept of mastery of the concept is comprehensively overcome by the features of the material that presents multimedia (text, mind map, audio image, video) and the assessment of material concepts. Features of material and assessment (both concept and practice of repetitive teaching skills) integrated into one tangible media based on Android applications. Mindmap is able to avoid misconceptions. The prospective teachers must avoid misconceptions in advance so that learning competencies can be achieved [30]. In the CK analysis, the material is adjusted to the lesson plan [31], this is because the learning media function to achieve the learning objectives planned for the lesson plan. The integration of technology that has been applied to micro-teaching include: (1) Network Multimedia Technology Reflected in the Microteaching for recording, storage, transmission, regulate and recurrence[32]; (2) video open class, (3) MOOC[33]; (4) video technology[10], [34]–[36]; and (5) use of ICT technology such as Laptops, LCD projectors, Desktops, EDUCOM, Smart classes, Memory sticks to find material sources[37]. The other novelty in this paper is the strategy of using smartphones as a teaching aid platform for micro-teaching. The smartphone usage strategy in question can be seen in the Table 2.

Table 2. Smartphone Strategies and Contributions

Strategies (features)	Contribution
Content material	mastery of concepts, especially with the presence of mind maps, examples of teaching skills and teaching practice videos
Questions	assessment of the results of mastering the concept
Chatbot	mastery of concepts through strategies to explore the ability to ask questions
Verbal teaching skills practice	practice verbal skills in repetitive teaching skills
Progress	to monitor the learning activities that have been done by users on the application
Notification	to notify users about learning activities that need to be done

The results of TPACK's analysis and strategy features on the smartphone are used to determine the specifications of the device needs. Product specifications developed are Android-based applications that are able to provide interactive material, display interactive dialogs or chats between users and bots, test verbal skills through recorded oral tests and display progress. Hardware requirements include (1) Android with a Lollipop operating system; (2) i5 processor specification laptop with 4GB RAM; and (3) Device specifications for 2GB RAM and Lollipop operating system. Software Requirements are (1) Android Studio with version 3.0.1; and (2) Android SDK with API 26. Functional requirements for applications that are capable of displaying: (1) splash screen; (2) showcase; (3) messaging display; (4) expenses when users enter the word; (5) material that functions as keyword messaging; (6) the progress of user value; and (7) license and credit menus. Menu needs consisting of: (1) Material containing a collection of materials related to microteaching concepts presented in the form of text, audio and video; (2) Problem Training containing questions that can be done to test knowledge about concepts that have been learned related to microteaching; (3) Conversation (chatbot), where users can have conversations with bots to deepen aspects and items of teaching skills; and (4) the practice of teaching skills (verbal ability - speech recognition) to test the user's verbal skills through the user's oral pronunciation related to eight teaching skills. The work system of the application can be shown in Figure 1.

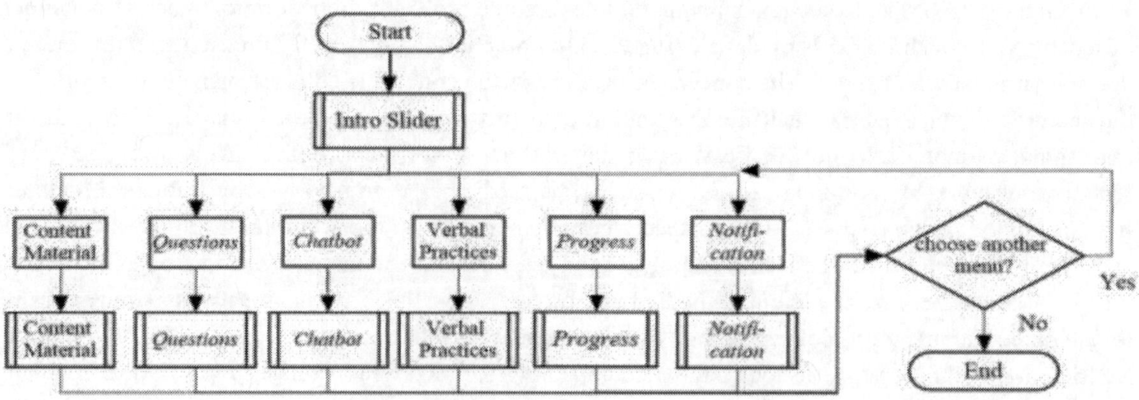

Figure 1. Application flowchart

When the application is first to run, there is an intro feature that displays the features that are in the application briefly. Furthermore, users can make menu selections as desired. Users can choose other menus after opening a menu. Videos on material features are made based on storyboards that refer to the assessment sheet on aspects of teaching skills. There are 9 videos consisting of teaching skills, namely, namely: (1a) learning opening skills; (1b) skills to close learning, (2) skills to explain; (3) questioning skills; (4) skills to make variations; (5) skills to provide reinforcement; (6) class management skills; (7) teaching skills for small groups and individuals; and (8) skills in leading small group discussions. Design results in the form of data design, architectural design, interface design, and algorithm design. The development stage consists of coding and testing errors in the program code created. The evaluation phase includes three things, namely control structure testing, functionality testing, and usability testing. The evaluation phase in the form of control structure testing uses testing on android developers, where all get the comments "process finished with exit code 0". Figure 2 shows an example of the results of the unit test in the progress and notification feature, where the top left of the test results progress feature, the bottom left of the notification feature testing feature, and the right part shows the progress and notification display features.

Figure 2. Example Unit Test Results (progress and notification features)

The evaluation phase in the form of black box testing to perform functionality tests shows that 100% of the devices have successfully shown the targeted function. There is a suggestion from the expert who performs the functionality test that is to add navigation "back" or return to the test menu (work on the question) and need to add the question number to the test menu. Both of these suggestions are followed up by adding back navigation and giving a test number. Media expert validation was done by two media experts. This validation was used to test the feasibility of fulfilling the elements of learning media. The result of the score is 3.79 which shows that from the media aspect, then "Microteaching" is suitable to be used as a learning medium. Validation of material experts is done to two material experts. The result of the 4.07 score shows that from the material aspect, "Microteaching" contains appropriate material regarding aspects of teaching skills. Test users to lecturers and students to conduct usability tests. The result of the acquisition is 3.91 which shows that from "Microteaching" is worthy of use. Figure 3 shows that all aspects of each evaluation show a score above 3.4 with "decent" interpretation. For aspects of benefit in the evaluation of material experts and technical aspects in the evaluation of media experts show very decent. The assessment results show that the material aspects of the media are in accordance with the lesson plan and technically, the media is easy to use. Even though the media received a proper assessment from each evaluation, the development team continued to make improvements based on the input obtained from the media assessment. From various improvement suggestions, Table 2 shows some improvements that have been made. Some display results from the Android application developed in this study can be seen in Figure 4.

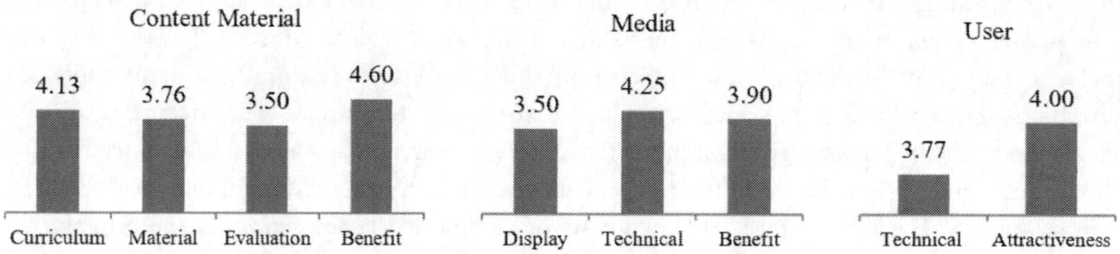

Figure 3. Test results of media experts, material experts, and users

Table 3. Revisions

No.	Suggestions for improvement	Results
1.	The icon is not clear	Change the main menu icon to make it clearer
2.	The material menu needs to be made more interesting	Add images and mind maps to the material to make it interesting.
3.	On the test menu, users cannot return to the previous question.	Add a reminder to avoid mistakes in selecting answers.
4.	In working on the test questions, given the number of the questions and the number of questions.	Add information about the problem number and number of questions.
5.	Some illustrations are given for each material to make it more interesting and not monotonous.	Add illustrations to the material

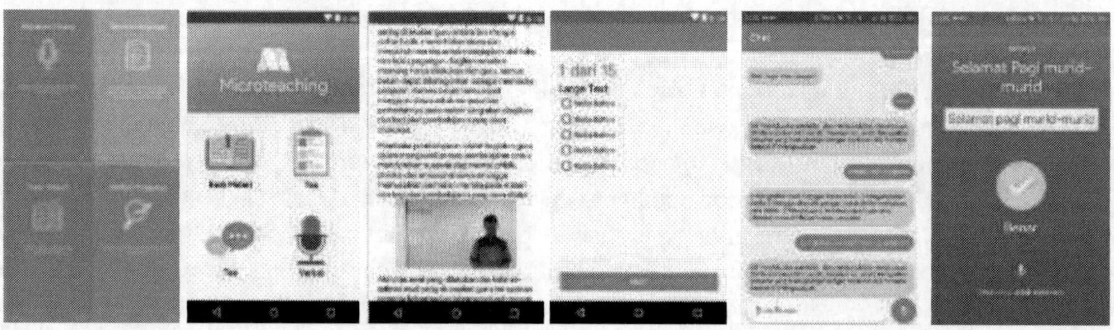

Figure 4. Some display results from the Android application developed in this study (intro, features, content material, questions, chatbot, verbal practices)

Application functionality and assessment of feasibility in each aspect of evaluation that is considered feasible, indicating the media can be used in learning. Because of that, "Microteaching" fulfills the elements of learning media, among others, the message to be transferred to students can be conveyed well, learning objectives can be achieved, helping students in understanding the concept of material, improving and improving learning processes and outcomes, and increasing student motivation in learning. The advantage of the application developed is the existence of mind maps to support the mastery of concepts and the existence of examples of teaching practices in the form of videos. The video is easier for students to get concrete examples of teaching practice. The chatbot feature is a new method offered by the team to accommodate and stimulate students' questioning skills. Efforts to improve skills, especially the ability to ask questions for prospective teacher students are needed[38]. The results of the needs analysis that cannot be implemented in this application are that the chatbot feature is still not comprehensive and still limited to displaying the material and the teaching skill practices feature is still limited to the practice of speaking, without assessing good teaching gesture. For further research, it is expected to be able to add material and interactivity to the chatbot feature, while the teaching skill practices feature adds a gesture assessment feature to teaching skills especially for non-verbal assessment, it needs to be developed to be able to assess non-verbal ability (gesture) by utilizing the development of computer vision (gesture recognition) and affective computing while still integrating the smartphone platform to facilitate users.

3. Conclusion

Technological involvement in learning activities cannot be avoided, technology integration in micro-teaching is one of the efforts to face these challenges. From various tests and revisions, the results show that: (1) the android-based application is used in learning with the bank features of the question bank, chatbot, audio interaction through the application of hybrid models (tutorials and drills) for understanding the concept of general microteaching material assessed through tests multiple choice and repeated training and assessment of teaching skills practices with menu display (a) demo; b) material bank; (c) chatbot; (d) multiple choice tests; (e) teaching skill practices; (f) progress; (g) notifications and licensing and credit menus; and (2) evaluation results show that control structure testing using testing on android developers shows that all codes have been tested by the unit and indicate that the code has passed all tests and there is no error in the code; (2) test functionality worth 100% indicates that the application can be run without any constraints; (3) the validation of media experts is 3.79 and the material expert is 4.07, both of which indicate that they are suitable to be used, so that the "Microteaching" application can be said to have fulfilled the elements of the learning media; (4) test the user to do usability test with a value of 3.91 indicating that the "Microteaching" application is technically easy and interesting to use; and (5) product revisions (applications) have been carried out. There are limitations in this case the chat feature is still not much and is still limited to displaying material. Nevertheless, the application has the advantage of having complete material accompanied by mind maps and videos. The results of the development of this application, are considered capable of realizing a media that integrates material concepts, examples and limited practices.

4. References

[1] Sukaswanto, "Mengefektifkan Peran Mahasiswa Peserta Pengajaran Mikro dalam Mensupervisi Pelaksanaan Pengaaran Mikro," *J. Pendidik. Teknol. dan Kejuru.*, vol. 21, no. 3, pp. 238–245, 2013.

[2] Helmiati, *Micro Teaching: Melatih Keterampilan Dasar Mengajar*. Yogyakarta: Aswaja Pressindo, 2013.

[3] A. Remesh, "Microteaching, an Efficient Technique for Learning Effective Teaching," *J. Res. Med. Sci.*, vol. 18, no. 2, pp. 158–163, 2013.

[4] F. R. Choudhary, M. R. Choudhary, and S. K. Malik, "Application of Microteaching in Teacher Education Prorgrams: A Meta-Analysis," *Lang. India*, vol. 13, no. 2, pp. 69–91, 2013.

[5] R. Koross, "Micro teaching an efficient technique for learning effective teaching skills: Pre-service teachers' perspective," *IRA-International J. Educ. Multidiscip. Stud.*, vol. 4, no. 2, pp. 289–299, 2016.
[6] Sukarti, *The application of student teaching skills is the practice of teaching in schools according to the tutor teacher in partner schools*. 2018.
[7] S. Al Darwish and A. Sadeqi, "Microteaching impact on Student Teacher's Performance: A Case Study from Kuwait," *J. Educ. Train. Stud.*, vol. 4, no. 8, pp. 126–134, 2016.
[8] T. Fakultas, "RPS Pembelajaran Mikro," Yogyakarta, 2017.
[9] E. S. Sharon, L. L. Deborah, and J. D. Russell, *Instructional Technology and Media for Learning, 10th Edition*. New Jersey: Pearson, 2012.
[10] T. J. Diana, "Microteaching Revisited: Using Technology to Enhance the Professional Development of Pre-Service Teachers," *Clear. House A J. Educ. Strateg. Issues Ideas*, vol. 86, no. 4, pp. 150–154, 2013.
[11] H. Fadhl, D. Halina, and A. Basheer, "Usability Guidelines of Mobile Learning Application," *J. Inf. Syst. Res. Innov.*, vol. 5, no. special issue, p. pp 70-77, 2013.
[12] G. Woodill, *The Mobile Learning Edge: Tools and Technologies for Developing Your Teams*. New York: McGraw-Hill Education, 2010.
[13] M. Ally, *Mobile Learning Transforming the Delivery of Education and Training*. Edmonton: AU Press, 2009.
[14] N. A. Ismail, F. Ahmad, N. A. Kamaruddin, and R. Ibrahim, "A Review on Usability Issues in Mobile Applications," *IOSR J. Mob. Comput. Appl.*, vol. 3, no. 3, pp. 47–52, 2016.
[15] Statista, "Number of smartphone users worldwide from 2014 to 2020 (in billions)," 2018. [Online]. Available: https://www.statista.com/statistics/330695/number-of-smartphone-users-worldwide/. [Accessed: 05-Jun-2018].
[16] W. U. N. Zuhra, "Kelahiran Generasi Z, Kematian Media Cetak," 2017. [Online]. Available: https://tirto.id/kelahiran-generasi-z-kematian-media-cetak-ctLa. [Accessed: 05-Jun-2018].
[17] M. B. Triyono and P. Utami, *Pembelajaran e-learning pendidikan vokasi*. Yogyakarta: UNY Press, 2017.
[18] Statista, "Global mobile OS market share in sales to end users from 1st quarter 2009 to 2nd quarter 2018," 2018. [Online]. Available: https://www.statista.com/statistics/266136/global-market-share-held-by-smartphone-operating-systems/. [Accessed: 05-Jun-2018].
[19] J. E. Istiyanto, *Pemograman SmartPhone menggunakan SDK Android dan Hacking Android*. Yogyakarta: Graha Ilmu, 2013.
[20] N. S. H, *Pemrograman Aplikasi Mobile Smartphone dan Tablet PC Berbasis Android*. Bandung: Informatika, 2012.
[21] Y. Miarso, *Menyemai Benih Teknologi Pendidikan*. Jakarta: Kencana Prenada Media group atas kerjasama Pustekkom-DIKNAS, 2009.
[22] A. Arsyad, *Media Pembelajaran*. Jakarta: PT Raja Grafindo Persada, 2007.
[23] S. Mulyanta and M. Leong, *Tutorial Membangun Multimedia Interaktif-Media Pembelajaran*. Yogyakarta: Universitas Atma Jaya Yogyakarta, 2009.
[24] B. Igler, T. Braumann, and S. Böhm, "Evaluating the Usability of Mobile Applications," *Int. J. Bus. Manag. Stud.*, vol. 5, no. 1, pp. 92–102, 2013.
[25] P. Kortum and M. Sorber, "Measuring the Usability of Mobile Applications for Phones and Tablets," *Int. J. Hum. Comput. Interact.*, vol. 31, no. 8, pp. 518–529, 2015.
[26] R. S. Pressman, *Software Engineering A Practitioner's Approach Seventh Edition*. New York: Mc Graw Hill Higer Education, 2010.
[27] D. Rodgers, "The TPACK Framework Explained (With Classroom Examples)," 2018. [Online]. Available: https://www.schoology.com/blog/tpack-framework-explained. [Accessed: 05-Jun-2018].
[28] N. Sudjana, *Penilaian Hasil Proses Belajar Mengajar*. Bandung: PT Remaja Rosdakarya, 2016.
[29] G. Zhou, J. Xu, and D. Martinovic, "Developing pre-service teachers' capacity in teaching

science with technology through microteaching lesson study approach," *Eurasia J. Math. Sci. Technol. Educ.*, vol. 13, no. 1, pp. 85–103, 2017.

[30] R. D. Setyawati and I. Indiati, "Analysis misconception of integers in microteaching activities A," in *4th International Seminar of Mathematics, Science and Computer Science Education, IOP Conf. Series: Journal of Physics: Conf. Series 1013*, 2018.

[31] G. P. Cikarge and P. Utami, "Analisis dan Desain Media PEmbelajaran Praktik Teknik Digital sesuai RPS," *ELINVO(Electronics, Informatics, Vocat. Educ.*, vol. 3, no. 1, pp. 92–105, 2018.

[32] Y. Jiugen, Z. WEnting, and W. Qi, "Microteaching Based on Internet and Multimedia Technology," in *The 8th International Conference on Computer Science & Education*, 2013, no. Iccse, pp. 885–887.

[33] Y. Zhang, "Strategies for Design and Application of Micro-lesson," *IEEE 2014 Int. Conf. Educ. Innov. through Technol. Strateg.*, pp. 85–87, 2014.

[34] E. O. Ekpo-Eloma, A. Arikpo, and N. E. Catherine, "Integrating video Technology in Micro-Teaching Sessions for Teacher-Trainees' Self-Appraisal And Professional Growth," *Glob. J. Comput. Sci. Tecnol. Interdiscip.*, vol. 13, no. 4, pp. 24–28, 2013.

[35] V. Can, "A microteaching application on a teaching practice course," *Cypriot J. Educ. Sci.*, vol. 4, pp. 125–140, 2009.

[36] A. T. Karçkay and Ş. Sanli, "The effect of micro teaching application on the preservice teachers'teacher competency levels," *Procedia - Soc. Behav. Sci.*, vol. 1, pp. 844–847, 2009.

[37] B. Bhattacharjee and K. Deb, "Role of ICT in 21 st Century's Teacher Education," *Int. J. Educ. Inf. Stud.*, vol. 6, no. 1, pp. 1–6, 2016.

[38] Wiyanto, S. E. Nugroho, and Hartono, "The Scientific Approach Learning : How prospective science teachers understand about questioning," in *The 3rd International Conference on Mathematics, Science and Education 2016, IOP Conf. Series: Journal of Physics: Conf. Series 824 (2017)*, 2017.

IOP Conf. Series: Materials Science and Engineering **535** (2019) 012010 doi:10.1088/1757-899X/535/1/012010

The mind-mapping learning model in the robotics course

KI Ismara [1], A Asmara[2], H Sigit P[3] and R Asnawi[4]

Mechatronics Engineering Education of the Faculty of Engineering Yogyakarta State University, Yogyakarta, Indonesia

E-mail : imaismara@uny.ac.id

Abstract. This paper aims to describe the mind-mapping learning model employed in the Robotics course and the results of its learning process. Used the classroom action research, the study was conducted in four meetings in the Electrical Engineering Education Department of the Faculty of Engineering Yogyakarta State University (YSU). The research participants were 21 students of the Robotics course. The data were collected by questionnaires and tests, the pre- and post-test, and analyzed descriptively. The results stated that the mind-mapping learning model, which was delivered in form of videos, could increase students' understanding of the Robotics course with an average score of 3.6 out of the 4.0-Likert scale. The average post-test score indicating the students' mastery learning of Robotics was 92.

1. Introduction

Technology has developed in various fields predicted to advance rapidly. Used in these fields, the forward-looking technology includes the robotics that involves industrial and human assistant robotics, Internet of Think (IoT) built to assist people to keep their tasks on track, and Blimps involving the hot air balloon development for long-term transportation [1]. One of them, robotics, has been known to enter Indonesia's world of industry, automotive, and even education.

Robotics in the world of education has become either a course subject or a study program or expertise specializing in the science of Robotics. Robotics as a course concerns on the matters in the world of Robotics consisting of mechanics, electronics, and computers. These three topics are consequently delivered as a unity in the Robotics so that this course as a whole has a broad coverage of topics. Despite this broad coverage of topics, the knowledge they have is so interrelated that they cannot be separated from each other.

Whilst effective learning methods in Robotics are still on search, some attempts have been made to improve students' learning achievement. They comprise the use of multimedia, demonstration, and game. These are carried out in order that Robotics knowledge can be understood well by the students and then they are used to support the technology development in Indonesia.

For teachers or educators, choosing an appropriate learning model is necessary. Some considerations on the students' learning needs and the target learning materials and sources are among those that should be accommodated so that a certain kind of learning model can be applied effectively and support the success of students' learning. Furthermore, teachers are expected to have the motivation and the spirit of renewal in the learning process he lives. Sardiman [2] states that competent educators are those who are able to manage the teaching and learning process. The term managing has a broad meaning concerning how educators master basic teaching skills, such as opening and closing the lessons, giving explanations, using various media, asking questions, and giving reinforcement, to name a few. It also deals with how educators apply strategies, teaching and learning theories, and create conducive learning.

In line with this, Marsh [3] adds that educators should possess teaching competencies, motivate their learners, create instructional models, manage their classes, maintain communication with their learners, as well as plan and evaluate the lessons. All of these support the success of educators in teaching. One of the current learning models that have been developed to improve students' understanding of science learning is the mind-mapping model. Mind-mapping is the way of developing thinking activities in all directions and capturing various thoughts in various angles. It develops divergent thinking and creative thinking. Mind-mapping, also well known as the concept map, is a very powerful tool for the organizational thinking which also resembles the easiest way to put information into the brain and retrieve that information when needed [4].

Concerning this, some of the general courses in the Electrical Engineering Education Department of Faculty of Engineering YSU have implemented this learning model. The results are believed to improve the students' understanding and memory on the information that had been conveyed. However, this model has never been applied in the courses Engineering Science. Thus, it is expected that this learning model can improve the students' ability to understand and store information taught in Robotics.

2. Research Method

This study was carried out in the Class A and Class D of the Control concentration in the Electrical Engineering Education Department Faculty of Engineering YSU. The participants were 21 students who were taught by one of the authors, Herlambang Sigit Pramono. This study used a Classroom Action Research (CAR) method with several stages in each cycle. (1) Plan - The first stage was planning that involved all researchers consisting of experts in the field of Robotics, multimedia, and mind-mapping to decide the course design with the mind-mapping learning model incorporated in, the learning material to cover, and the right technique in applying the model. (2) Do - the action or implementation stage was carried out in the classroom, with one model of an educator who delivered the material by the mind-mapping model. (3) Check - the last stage was the evaluation and reflection on the implementation of the learning process. The data were collected by means of questionnaires and tests and were analyzed descriptively.

3. Findings and Discussion

3.1. Findings

The planning stage was carried out through discussions for deciding the materials used in Robotics, ranging from the underlying discipline materials to the techniques to apply the mind-mapping model. It was planned that the course would be delivered mostly in the lecture delivery model, with the mind-mapping model for teaching the basics of Robotics. Using the help of Inspiration 9.0.3 software in the delivery, Robotics was used as the basis of the materials to be developed. The following is the presentation of the concept map or diagram of Robotics developed by using the software.

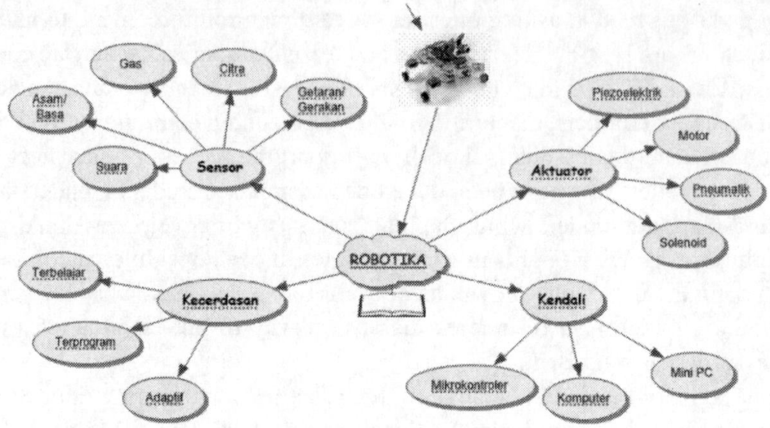

Figure 1. The Mind Map of Robotics

In the implementation stage, unexpectedly only a few students participated eagerly. There were only three active students answered and discussed the notions of Robotics. Nevertheless, the results of the post-test increased significantly compared to those of the pre-test. The following is an overview of the results of the pre- and post-test.

Table 1. The Students' Results of the Pre- and Post-test in Cycle 1

Student	Pre-test	Post-test
1	76	94
2	85	96
3	78	90
4	64	82
5	80	96
6	78	90
7	76	88
8	78	86
9	82	96
10	68	86
11	64	86
12	84	98
13	82	96
14	76	92
15	72	90
16	78	92
17	72	88
18	80	96
19	78	86
20	84	94
21	76	86
Mean	76.71	90.86

The improvement of the post-test scores shows that there was an increase in the students' understanding of the materials delivered. Meanwhile, it was evaluated that the students' participation was still low. Thus, in the next cycle, the researchers decided to improve their participation by using the actual instruments or tools of Robotics instead of the diagram. The next topic of the presentation materials was the actuator in robotics.

In the second meeting or the second cycle, unfortunately, the teaching and learning process got unsatisfactory results, as the students' participation was still low. However, there were six students participated actively during the lesson, indicating an increase in terms of the number of students participating in. The increase occurred because they were interested in the media that were brought so that they began to question or show the desire to find out more about the robot working principles and the procedures to operate the robot actuators.

The results post-test scores increased compared to those on the pre-test, with the pre-test mean score of 84.62 and 92.76 as that of the post-test. This improvement is because the material used was the exam blueprint on the robot actuators some of those have already given in the previous courses. The results of this second phase evaluation were used as a consideration to increase the participation of learners.

Then, on the planning stage for the third cycle, it was triggered that the students would be the ones who made the mind map of the determined topic. This idea was believed to be the solution to significantly increase the students' participation. In addition, it can enhance the students' cooperation in that they discussed the materials, shared ideas, and made decisions together with other students.

The third stage, consequently, was carried out by involving students to create a mind map of a set of the subject-related knowledge of Robotics. By the method of a group discussion and a determined deadline, students made a groove mind map about the Robotics Sensor. In this stage, students were divided into four groups with each 4 to 5 members and were assigned to make a concept map on the topic of Robotics Sensor on their own as provided in Figure 2.

The learning outcomes of this cycle show an increase in the students' participation, namely, they expressed their opinions, decided the form of the mind map, and asked something they did not know about the topic presented. The final product of this third cycle is the mind map on the Robotics Sensor itself. Although when viewed from the data the compatibility of results of the mind map with the answer key gain only 70%, the students' participation increased as planned, namely 21 active students participated in the group work and their learning interest was high.

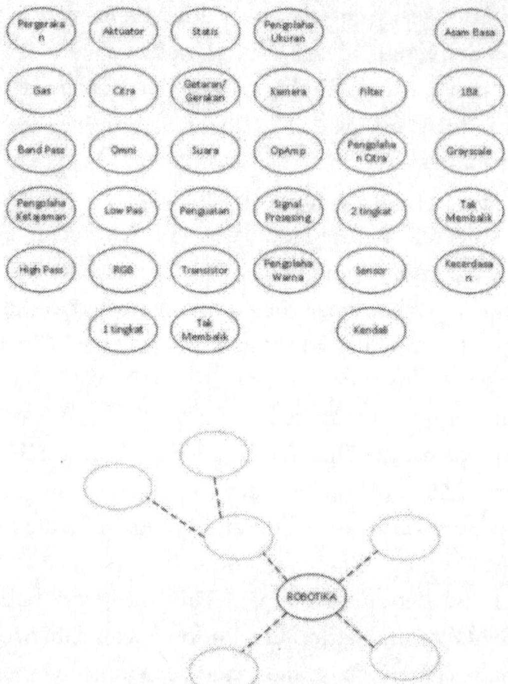

Figure 2. The Interactive Concept for the Mind Map of Robotics Sensor

The results of the third cycle were used as an evaluation to plan the fourth phase. In the fourth stage, the topic of Robotics Image Processing was delivered by the lecturing and student discussion. New terms in the topic were introduced to make the students more active in finding out the meaning of the new terms appeared. The evaluation was done at the beginning and end of the meeting to see if the results increased with the method applied.

The final results of the implementation of the fourth meeting of the learning process are very satisfactory, indicated by the pre-test mean score of 88.34 and the post-test mean score of 93.43. At the fourth meeting, questionnaires were also used to evaluate the students' comprehension of the subjects delivered by using the mind-mapping learning model. Of the 21 students, the average score obtained from the questionnaire is 3.6 out of the 4.0-Likert scale. The following is the summary of the questionnaire data from the 21 participants.

Table 2. Students' Comprehension Scores Collected through Questionnaires

Student	Scores
1	3.4
2	3.8
3	3.5
4	3.8
5	3.8
6	3.6
7	3.6
8	3.5
9	3.8
10	3.8

IOP Conf. Series: Materials Science and Engineering **535** (2019) 012010 doi:10.1088/1757-899X/535/1/012010

11	3.6
12	3.6
13	3.8
14	3.4
15	3.8
16	3.8
17	3.8
18	3.2
19	3.6
20	3.5
21	3
Mean	**3.60**

3.2. Discussion

The data obtained show that the mean score for the improvement on the students' understanding is 92.35, taken from the three post-test results that had been conducted. This shows that the students' understanding of the engineering lecture materials can be improved by the mind-mapping learning model. Meanwhile, to apply this learning model, educators should decide the potential topics that can be developed into concept maps. It is believed that arranged flow of thought presented in the mind maps will facilitate the understanding and memory of the students on the topics.

Furthermore, the average score of the students' understanding at the end of the meeting is 3.6 out of a 4.0-scale, demonstrated the two third of the students' understanding of the topics in the subjects, which in other words shows that they gained a good understanding of the learning materials. This improvement occurred because of the mind maps that were well organized, starting from the basic units of the maps to their branches. This certainly affects the students' ability to memorize challenging learning materials.

The mind-mapping learning model, thus, may be combined with other learning methods to improve the results of the learning process. For example, in this study, the learning model was combined with the lecture method, simulation/modeling, and discussion. Each combination should have certain goals to achieve in the learning process.

Based on the results of this study, some steps should be done to maximize the impacts of an effective mind-mapping learning model. They include the followings. (1) Educators (teachers or lecturers) tell the students the learning objectives of every topic. (2) Educators (teachers or lecturers) tell the students the learning methods and models they are going to apply. To gain effective learning process and results with the mind-mapping learning model, students' participation during learning is vital. (3) Educators (teachers or lecturers) divide the students into small groups consisting of 3-4 students. (4) Each group should work on a certain problem chosen from the determined topics with the introduced terminologies. (5) Students are required to discuss the problem to propose solutions by looking for information or the meaning of the terms from the given topics. (6) Each group should make a summary in the form of a mind map, and then the group members are chosen randomly to present their mind map. (7) Educators provide the answer key or the solutions to the problems discussed. (8) Students and educators come to kinds of shared conclusions on the learning topics.

Besides, this study has both products in the form of reports and articles and learning videos which can be used as one of the references in the implementation of the mind-mapping learning model. Videos were uploaded to a free video service provider so that readers can retrieve or download them for free.

IOP Conf. Series: Materials Science and Engineering **535** (2019) 012010 doi:10.1088/1757-899X/535/1/012010

4. Conclusion

From the research that has been done, some conclusions can be drawn. They are as follows. (1) The mind-mapping learning models can improve the students' understanding, with a mean score of 3.6 out of a 4.0-scale. (2) The results of the learning process delivered by the learning model attain a very good learning mastery with a mean a score of 92 taken from the three post-tests in three different cycles. (3) The output of this study can be used as one of the references in producing learning videos applying the mind-mapping learning models.

5. References

[1] Dom Einhorn, 2015. *The Future Of Mining The Internet Of Things Robotics And Blimps* (access at https://born2invest.com)
[2] Sardiman, A. M. 2004 *Interaksi dan Motivasi Belajar – Mengajar* (Jakartssa: Rajawali)
[3] Marsh, Colin. 1996. *Handbook for beginning teachers* (Sydney: Addison Wesley Longman Australia Pry Limited)
[4] Buzan, Tony 2008 *Buku Pintar Mind Map* (Jakarta: PT. Gramedia Pustaka Utama)

IOP Conf. Series: Materials Science and Engineering **535** (2019) 012011 doi:10.1088/1757-899X/535/1/012011

Improving the compact crucible furnace performance by adding fins in heating chamber

A Leman[1], Tiwan[2], Widarto[3], and A Ardian[4]

Mechanical Engineering Education Department, Yogyakarta State University, Yogyakarta 55281, Indonesia

E-mail: widarto@uny.ac.id

Abstract. A crucible furnace is developed continuously to obtain appropriate furnace in support casting practice at VHS. This research aimed at improving performance of compact crucible furnace by modification the heating chamber. Fins were placed at inner side of compact crucible furnace wall. The furnace performance determined by increasing of temperature rate to melt 3 kg of aluminium. Descriptive analysis is used to describe performance of modified compact crucible furnace. The fins addition on the inside wall improves the compact crucible furnace performance. The temperature rate is increase as the fins inclination angle increase. Fins with inclination angle of 80^0 could keep and maintain the temperature of heating chamber. Compact crucible furnace which added fin on the inside wall is able to melt 3 kg of aluminium in 35 minutes.

1. Introduction

Nowadays, most of ASEAN nations deal with the need of skilled labor [1]. This is directly related with vocational high school (VHS) as one of institution that provide skilled labor. The VHS should support the students in mastering competencies needed by stakeholders. One of that competence is casting process which substantial in product manufacturing. Regulation of Directorate General of Primary and Secondary Education of Indonesian Education and Culture Ministry number 06/D.DD5/KK/2018 dated 7 June 2018 about VHS expertise spectrum state that VHS student majoring in mechanical engineering should possess casting competency [2]. This indicate that is important for mechanical engineering of VHS student to learn casting process.

Furnace to melt metal is the main device in casting process. There are several kinds of furnace but the crucible furnace is the one that suitable for casting practice at VHS. The reason is the crucible furnace has simple construction thus easy to made, operate, maintain and low cost. The main features is the existence of crucible pot in the furnace on a support block. Charcoal, oil, or gas can be used as the energy source [3, 4, 5, 6]. There are two holes, one at the bottom side to supply heat from the energy source and the other at upper side to throw out combustion waste. Conventionally, the bottom hole is in the tangential direction. Thus, the heat will flow around crucible pot and then transferred to melt metal inside the pot.

The crucible furnace with liquid petroleum gas (LPG) as the energy source has been developed continuously [7, 8, 9, 10, 11]. The sustain development goal is to obtain appropriate furnace for VHS. The compact crucible furnace is the latest development at this time that appropriate for casting practice at VHS [9, 10]. It is being applied at several VHS. The advantages of compact crucible furnace are compact dimension (450x450x900 mm), simple construction, capable to melt 3 kg aluminum in 40-50 minutes, and the average gas consumption 2 kg. Unlike the conventional, the bottom hole in the compact crucible furnace placed exactly under the crucible pot.

Furthermore, the development also aimed at increasing the energy efficiency. Tanaka et.al. [12] had been developed a guide way from cast refractory in the inner side wall of crucible furnace. The purposes is to make sure the heat flows around the crucible pot perfectly. The heat flow like passing the screw line so it will contact longer with the pot. This technique could improve the furnace efficiency. Another method introduced by Arianto et.al. [11]. Fin from steel welded to the pot so that form a screw line. Like Tanaka's method, the heat will follow the fin and flow perfectly around the pot. Moreover, the steel fin will catch the heat as it heated up. The heat then transferred conductively to melt the metal inside pot. Both development above was conduct for conventional crucible furnace.

This article will discuss an addition of steel fin in compact crucible furnace heating chamber to improve the performance. The concept is to hold the generated heat stay longer in the heating chamber. The heat caught by steel fin, which will have heated up thus the heating chamber temperature raise quicker and heat will remain longer.

2. Method

The compact crucible furnace uses high pressure stove to burn LPG. The crucible pot made from steel pipe with outside diameter 115 mm, thickness 8 mm, and height 195 mm. A 10 mm thick of steel plate welded to the bottom side of the pipe. Fins were made from 3 mm thick of steel as shown in Figure 1. The variation of inclination angle (α) of the fin are 0, 20, 40, 60, and 80 degrees. Fins placed in heaating chamber as shown in Figure 2.

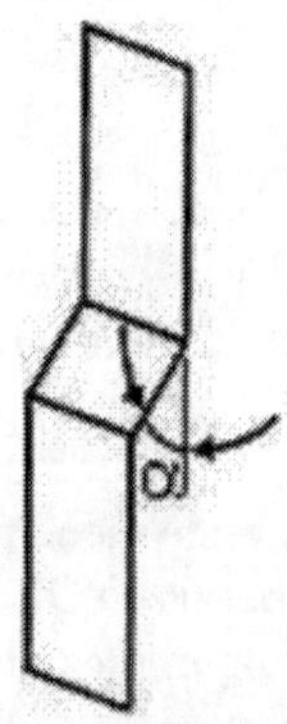

Figure 1. Fin from steel

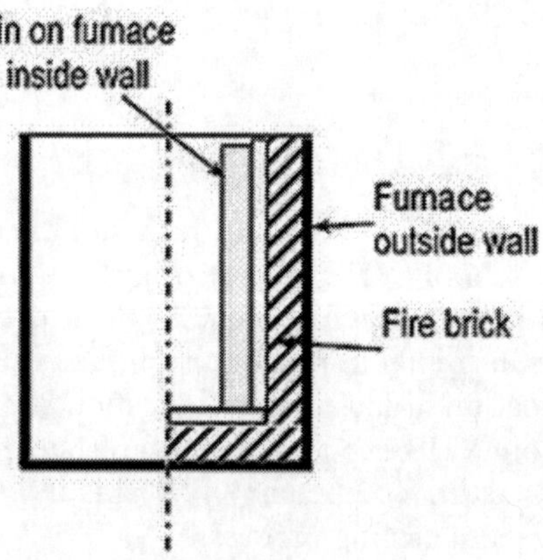

Figure 2. Fin placement in heating chamber

The experimental procedure was done with steps as follows: (1) Weighing 3 kg of aluminium, melted, and let it solidify in the crucible pot; (2) Placed fins onto the inner side wall of compact crucible furnace and warmed up the heating chamber up to 70-80 ^{0}C; (3) Weighing the initial weight of LPG using digital scale Slimline Electronic Scale TIF 9010A; (3) Heating the aluminium that had been prepared in the first step until it melt; (4) The temperature changes of aluminium, fins and compact crucible furnace inside wall were recorded every 5 minutes using infrared thermometer Krisbow KW06-409 until the aluminium melted. Repeat the steps above for other fins variation.

3. Result and Discussion

The compact crucible furnace with high pressure stove shown in Figure 3. The high pressure stove is placed at the bottom side thus the compact shape of the furnace achieved. The furnace made from rolled and welded of 2 mm thickness of steel plate [10]. The inner side furnace isolated by firebrick that installed using fire clay. The furnace also equipped with extended and deflected exhaust pipe as economizer [8]. The compact crucible furnace supported by a frame made from L profile steel of 40x40x3 mm.

The fin shape has design as shown in Figure 4. It welded on a steel plate that has two hooks. It has incline portion to directing the flow of heat in heating chamber of furnace. The incline portion is on lower side exactly at the crucible pot position. Inclination of fin were varying of 0, 20, 40, 60, and 80 degrees. The fins arranged in the heating chamber as shown in Figure 4. It arranged encircling the crucible pot. Fins deflect flow of the heat from stove to extend the path of the flow of heat so the heat will held longer in the heating chamber. Thus, when fins heating up as hit by flow of heat from stove below, the heating chamber temperature increase faster. Convection heat transfer occurs on the surface of crucible pot wall and then heat transferred through wall of crucible pot conductionally to the aluminium.

Figure 3. Compact crucible furnace with economizer

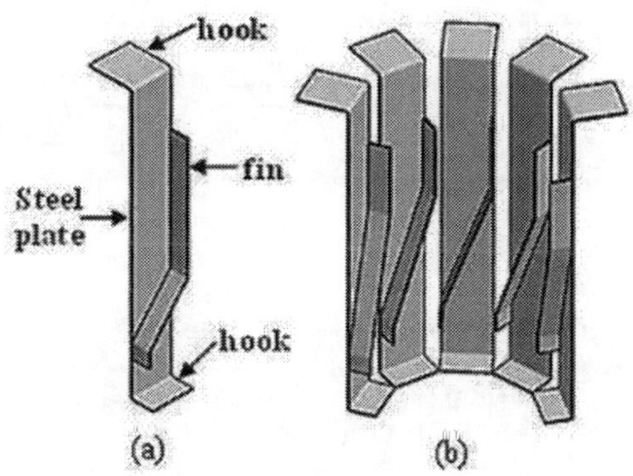

Figure 4. (a) Fin design; (b) Fin arrangement

Figure 5. Fins placement in heating chamber

Figure 6. Inclined portion of fins in red color

IOP Conf. Series: Materials Science and Engineering **535** (2019) 012011 doi:10.1088/1757-899X/535/1/012011

Figure 5 shows the placement of fins in the heating chamber. Fins were placed encircling pot on the compact crucible furnace hooked at inside wall. Figure 6 shows heating process in the furnace heating chamber. It can be seen that at the section where fins lie, the heating chamber more red in color. Further observation indicates that red color is concentrated at where fins inclined. It is proven that fins also isolating heat to stay longer around pot. Thus, heat around inclined portion of fins increase pot temperature faster. This condition occurs on all of variation of fin angle inclination.

Fins addition on heating chamber wall could improve compact crucible furnace performance as shown in Figure 7. It can be seen that temperature of fins increases faster compare to wall of the inside furnace and aluminium in the pot. This indicates that fins of steel absorb the heat. The heat then radiates to crucible pot thus both pot and aluminium inside pot achieve almost equal temperature. Furnace with inclination angle of fins of 0 degree able to achieve 650 ^{0}C in 50 minutes. This result similar to furnace development conduct by Arianto, et.al [10]. However, figure 7 shows that addition of fins will affecting the performance of compact crucible furnace. The increasing of inclination angle would give faster temperature increasing. The variable of inclination angle does not relate with variable of fin gap, which differ with study conduct, by Arianto et, al [10]. However, inclination angle more than 90 degree will causing flow of the heat turn back, which can cause damage to the stove. Furthermore, the curve slope is increase rapidly at first 5-10 minutes by the increasing of inclination angle of fin. It can be observed from Figure 7. This due to the ability of fin in retarding the heat flow and explain how pot temperature rise faster.

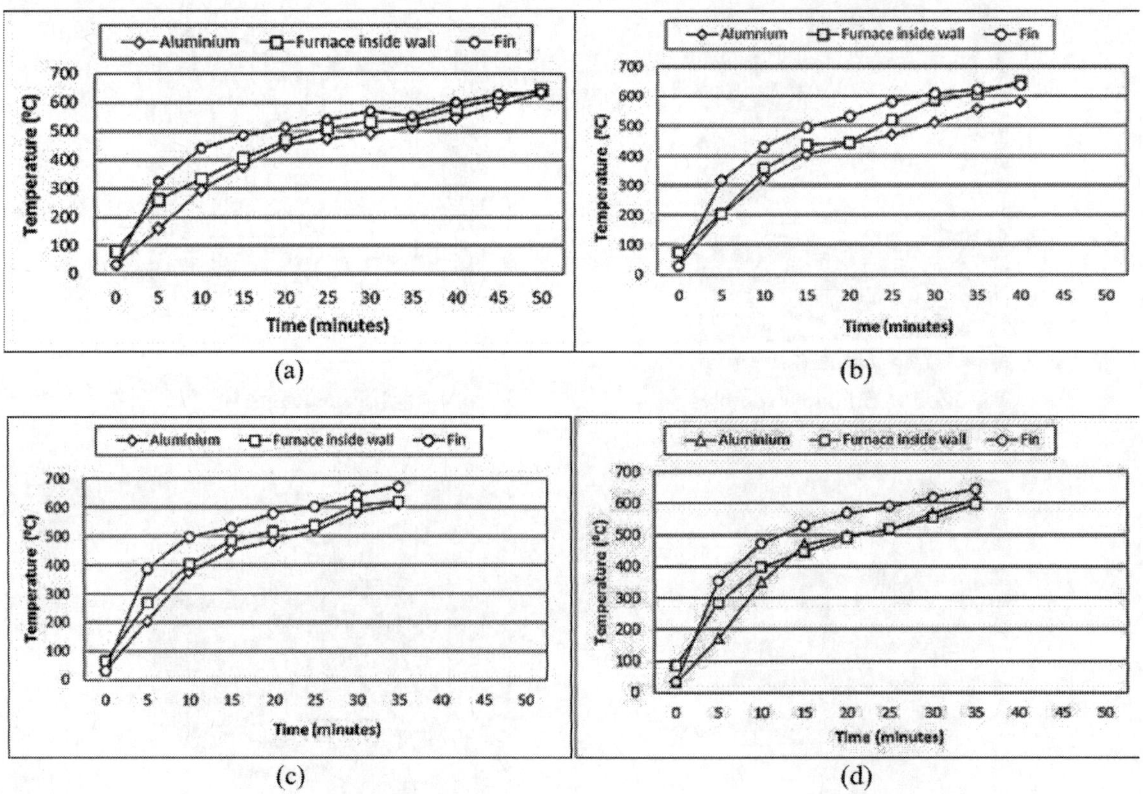

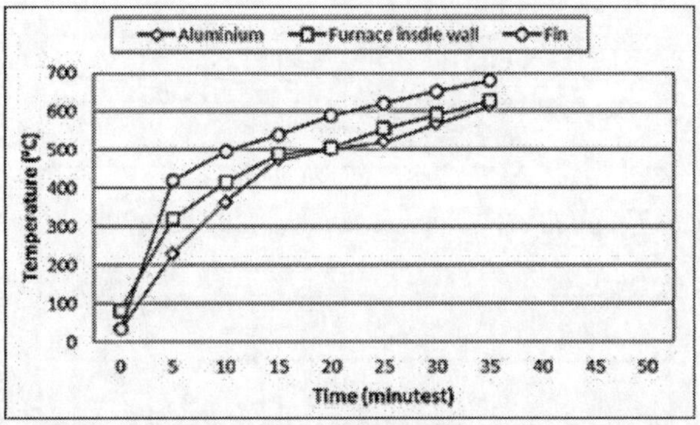

(e)

Figure 7. Effect of fin addition on performance of compact crucible furnace. Fin inclination angle: (a) 0^0; (b) 20^0; (c) 40^0; (d) 60^0; (e) 80^0.

At Figure 7e is observed that fin reach 680 ^{0}C in 35 minutes as well as the increasing of aluminium temperature. This indicate that flow of heat from below furnace restrained at inclined portion of fin which heat up quickly and then spread around crucible pot. After that, heat transferred by crucible pot wall of steel to aluminium.

Figure 8a shows similar tendency of temperature changes of aluminium on fin inclination angle of 40^0, 60^0, and 80^0. This due to the same heat transfer mechanism. However, slightly different tendency shown at Figure 8b. Heat-increasing rate on fin inclination angle of 80^0 is faster since heat restrained longer. Heat flow from below extremely deflect on fin inclination angle of 80^0. Thus, fin heated up quicker. Heat received by crucible pot and aluminium occur through conduction mechanism thus aluminium heat increasing rate is not too affected by fin inclination. Contrary with spiral finned crucible pot [11], the fin inclination angle and pot diameter determine fin pitch thus, there is optimum condition. In this study, fin hooked at inside wall of the compact crucible furnace so the crucible pot diameter is not take effect. It is been imply that fin inclination angle of 80^0 able to isolating heat at inclined portion better. This means, it will maintain temperature at this area when stove turned off and pouring process conducted.

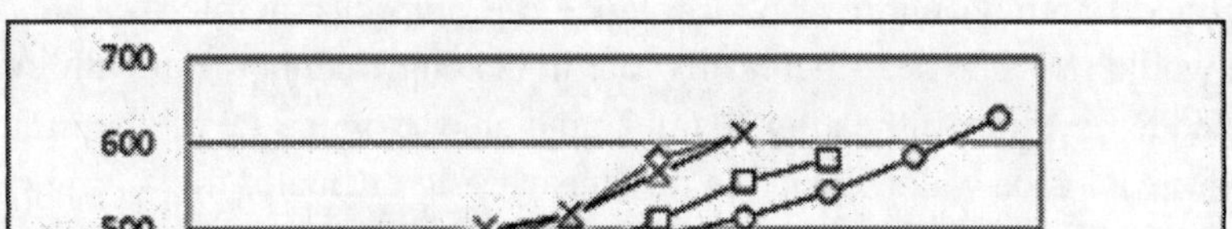

(a)

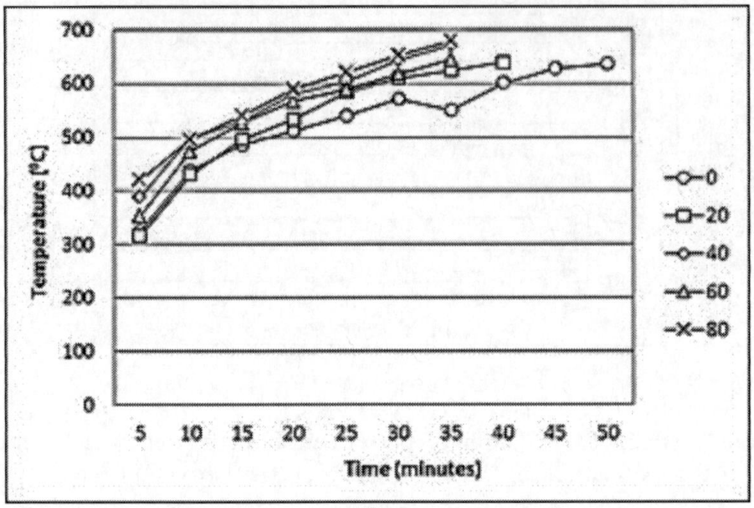

(b)

Figure 8. Effect of fin inclination angle on heat increasing rate of: (a) fin; (b) aluminium

However, there is an obstacle in this innovation. The compact crucible furnace has narrow heating chamber that limiting the use of crucible pot with wide diameter. Thus, when molten aluminium ready to poured, the crucible pot should be taking out because ladle cannot be use to pick up molten aluminium. Fin addition on inner sidewall of compact crucible furnace causes difficulty since space of heating chamber become narrower. Likewise, when crucible pot will put in again into heating chamber for next melting process. Further innovation must be conduct to improve the performance of compact crucible furnace with better operating. Heat requirements for melting aluminium is also not calculated. This is because the study purpose is only to determine the effect of fin additions and inclination angle on performance of compact crucible furnace. However, the heat requirement should be calculated to determine the furnace efficiency as fins added.

4. Conclusion

The using of inclined fin could improve performance of compact crucible furnace. Increasing of fin inclination angle would raise the heat-increasing rate in heating chamber. Fin with inclination angle of 80^0 could restrained temperature of heating chamber and able to melt 3 kg of aluminium in 35 minutes. However, hooked fin at inside wall of crucible furnace causing difficulty in pouring process.

5. References

[1] Gross M 2017 Developing a competency standard for TVET teacher education in ASEAN countries *Jurnal Pendidikan Teknologi dan Kejuruan* **23** 279-87

[2] Directorate General of Primary and Secondary Education 2018 Regulation number 06/D.DD5/KK/2018 (Jakarta: Indonesian Education and Culture Ministry)

[3] Aditya Nag M V and Dareddy R R 2012 Performance improvement of an oil fired furnace through oscillating combustion technology *Int J. Sci. & Eng. Research* IJSER(2012)

[4] Ighodalo O A, Akue G, Enaboifo E and Oyedoh J 2011 Performance evaluation of the local charcoal-fired furnace for recycling aluminium *J. Emerging Trends in Eng. and Appl. Sci.* **2** 448-50

[5] Olalerel A A, Dahunsi O A, Akintunde M A and Tanimola M O 2015 Development of a crucible furnace fired with spent engine oil using locally sourced materials *Int J. Innovation and Appl. Studies* **2** 281-88
[6] Suresh R and Nagarjun M G 2016 Construction and Performance analysis of Pit furnace by using Biodiesel *Indian J. Sci. and Tech.* IJST
[7] Soemowiagdo A L, Tiwan and Mujiyono 2014 Pengembangan tungku peleburan aluminium untuk mendukung kompetensi pengecoran di SMK program studi keahlian teknik mesin *Inotek* **18** 80-94
[8] Soemowidagdo A L, Tiwan, and Mujiyono 2017 Crucible furnace with economizer for casting practice at mechanical engineering education department faculty of engineering yogyakarta state university *Jurnal Dinamika Vokasional Teknik* **2** 21-27
[9] Ilyas M N and Soemowiagdo A L 2017 Compact crucible furnace for aluminium casting practice at muhammadiyah 1 salam VHS *Jurnal Dinamika Vokasional Teknik Mesin* **5** 9-19
[10] Soemowiagdo A L, Tiwan, Sutopo and Nurdjito 2017 Sarana Praktik Pengecoran Aluminium Untuk SMK Program Keahlian Teknik Mesin, *Prosiding Seminar Nasional Pendidikan Vokasional (Yogyakarta)* vol 1 (Yogyakarta: Faculty of Engineering/Yogyakarta State University) p 277-284
[11] Soemowidagdo A L, Tiwan, Widarto and Arddian A 2018 Spiral finned crucible pot *Proc. 3rd. Int. Conf. on Industrial, Mechanical, Electrical, and Chemical Engineering (Surakarta)* vol 1931 (America: A N Agung *et al*/ AIP) p 030066-1
[12] Tanaka A, Tsuri M, Ando H, Kikura M and Ueki M 2001 High efficeincy crucible melting furnace for non-ferrous metals (accessed 10 August 2017) from: http://www.h-techno.com/en/seihin/asm.html

IOP Conf. Series: Materials Science and Engineering **535** (2019) 012012 doi:10.1088/1757-899X/535/1/012012

Developing a control system for a racket-wielding robot using a combination of promoximity sensor and pneumatic actuator to optimize shuttlecocck hit

M Khairudin[1], and Soeharto[2]

Department of Electrical Engineering Education, Yogyakarta State University, Yogyakarta, Indonesia

E-mail: moh_khairudin@uny.ac.id

Abstract. This study aims at developing a stroke control system in a racket-wielding robot using a combination of proximity sensor and pneumatic actuator to optimize a shuttlecock hit. As the shuttlecock can come at any time and from any directions, this racket-wielding robot is developed to produce accuracy in its serving distance, service timing, and arm swing, resulting in an optimum hit. A research and development method was employed in this study with every stage of the sub-system development was tested for evaluation and revision. This study has developed and implemented a prototype of a stroke control system for a racket-wielding robot on the bases of proximity sensor and pneumatic actuator. The shuttlecock-wielding arm has functioned properly with a success rate of 100%, while the achieved hitting rate was 70%.

1. Introduction

A robot arm (also referred to as a robot manipulator) is a type of robot that is designed to carry certain load. As a payload, this robot offers more advantages compared to a conveyor does: it can lift a payload to different places, it is made of thin material, it is lighter, more energy-efficient, as well as easier to operate and manufacture, and it only requires a small actuator [1].

Features of the easiness in operating this robot arm include not only the time and distance accuracy but also the ability to carry or hit object until it reaches the designated area. This accuracy relies heavily on the type of sensor and actuator used. Therefore, a careful analysis and selection of the right type of these two elements are vital to produce the desired moves.

In developing a robot, especially a robot arm, the use of an accurate sensor for the right usage is very important. The sensor needed by a robot arm is the one that can reach, touch, or encounter objects accurately.

Other than a sensor that functions as the move controller against the presence of nearby objects, a robot needs an actuator/motor. A wrong choice in selecting the actuator type may cause the robot arm to have problems in moving to reach its target. An electric actuator is considered inappropriate to be used as an actuator for a racket-wielding robot. That is why in this study the researchers focus on the use of pneumatic actuator in a racket-wielding robot.

Employing a robot arm for moving things from one place to another is more advantageous than employing a conveyor. A robot arm can be used in many applications starting from goods delivery, microsurgery, and nuclear maintenance to space robotics [2]. The most challenging problem is related to the control system of the robot arm, specifically the long-distance control system, considering that a robot arm is often used in a danger area.

IOP Conf. Series: Materials Science and Engineering **535** (2019) 012012 doi:10.1088/1757-899X/535/1/012012

2. Result and Discussion

For practical applications, robot arms are chosen for they provide more flexibility in their applications so that the process of controlling and maintaining the accuracy of the arm position is quite challenging. For this purpose, it is very important to identify the flexibility nature of the thin material used through a mathematical model [3].

In calculating the characteristics of the robot arm's moves mathematically, Wang and Mills employ the Finite Element (FE) method to derive a dynamic model for vibration control in a one-link robot arm [4]. The flexibility of the one-link robot arm has already been calculated using the Particle Swarm Optimization algorithm [5]. Meanwhile, the flexibility and characteristics of a two-link robot arm are identified using the Assumed Mode Method [6]. Other researchers, Tian *et al.* have used the Absolute Nodal Coordinate method to predict the flexibility of a robot arm [7]. In addition, to anticipate over-flexibility, Olalla *et al.* have developed a robot arm controller using a strong control system [8].

An ultrasonic transducer is used to control a robot arm using the AT89C51 microcontroller. The distance between the arm and the targeted object is measured based on the time lapse in the detection of ultrasonic echo, resulting in a quite precise measurement. This method requires the operator to set a threshold value to obtain a minimum limit of the ultrasonic echo magnitude when the object is detected. This setting is done by a device called a potentiometer of which the value is often inconsistent due to the worn out sensor. When using this method, the microcontroller has to wait for the reflection wave to come. This waiting time is burdensome for the microcontroller, especially for microcontroller with heavy tasks like controlling a robot's move.

In order to decrease the dynamical equation of motion, the total energy related to the robot arm system must be calculated using a kinematic formulation [9].

This study focuses on designing a racket-wielding robot that can serve a shuttlecock to the opponent. The design of this robot is presented in Figure 1.

Figure 1. The Design of the Racket-Wielding Robot

The specifications of the developed racket-wielding robot are presented in Table 1.

Table 1. The Specifications of the Racket-Wielding Robot

Length	1100 mm
Width	1100 mm
Height	1400 mm
Materials	Aluminium profile, Acrylic, Iron, Nylon, Rubber
Battery	24 V ,8 AH

This racket-wielding robot is intentionally designed in a length, width, and height of 1100 mm, 1100 mm, and 1400 mm respectively to make the robot able to make a service. First, the shuttlecocks are stored in a shuttlecock tube. If the switch is turned on, the bottom lid of the tube will open and the shuttlecock will fall. When the shuttlecock is falling, it will be detected by the proximity sensor and thus trigger the racket actuator.

The data processing system in this racket-wielding robot utilizes the IC ATmega128 as both an ADC that can detect analog signal from a joystick and a PWM pulse generator. Later, the digital PWM signals are used to control the motor and the PWM servo controller. The voltage source used to supply the circuit and the motor is a lithium polymer battery with a capacity of 24 Volt, 8 AH.

The electronic circuit is attached to the robot body and is protected with a box made of an acrylic board. In that way, the circuit or the components inside the box will be safe whenever there is a shock caused by the robot's moves.

The Court. The court used to test this racket-wielding robot is an internationally standardized badminton court, exactly the same as that used in badminton by humans. The court used in this study is the one located in the former hall of Faculty of Engineering. Regarding the court dimensions, the length is 13.41 m and the width is 6.10 m. Meanwhile, the structure of the court is illustrated in Figure 2.

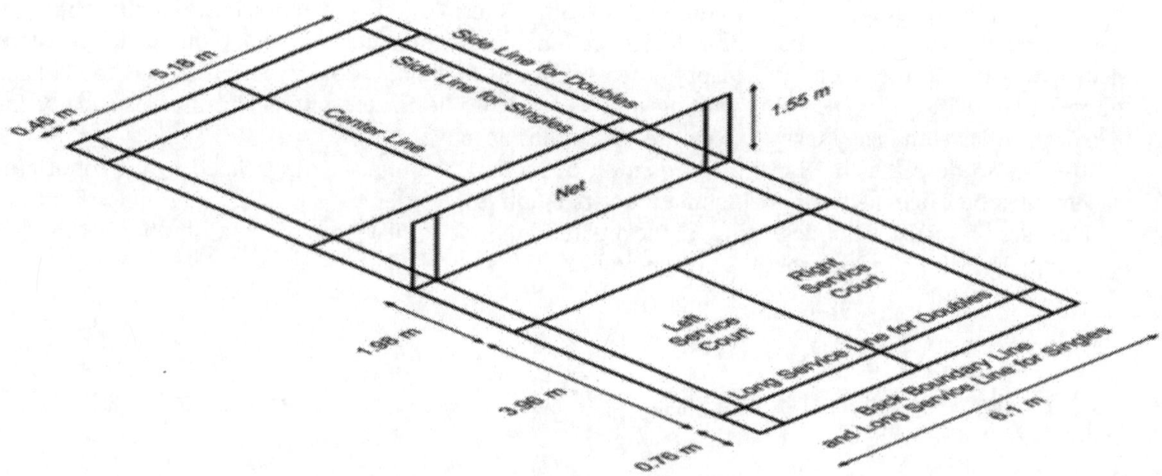

Figure 2. The Badminton Court Diagram

The green area is the playing space, the yellow area is the service area for the robot, and the gray one is the operator area. These three different colors are expected to help the robot make a service more specifically.

3. Conclusions

This study has developed a racket-wielding robot with manual functionality that is run by an operator. The swing made by the robot is that performed in serving a shuttlecock dropped by the racket-wielding arm with a pneumatic actuator. The forearm is used to make a low stroke, while the upper arm is for a high stroke. The shuttlecock-wielding arm has functioned properly with a success rate of 100%, while the success rate of the forearm in returning the opponent's stroke was only 70%.

4. References

[1] Khairudin M. Quantitative Feedback Theory-Based Robust Control For A Spindle Of Lathe Machine. International Journal on Smart Sensing and Intelligent Systems. Vol.9 No. 4, December 2016.

[2] Dwivedy, S. K. and Eberhard, P. 2006. Dynamic Analysis of Flexible Arms, a Literature Review. *Journal on Mechanism and Machine Theory*. 41(7): 749–777.

[3] Mohamed Z, M. Khairudin, A.R. Husain, and B. Subudhi. 2016. Linear matrix inequality-based robust proportional derivative control of a two-link flexible manipulator. Journal of Vibration and Control 22(5): 1244–1256.

[4] Wang, X. and Mills, J. K. 2005. FEM Dynamic Model for Active Vibration Control of Flexible Linkages and Its Application to a Planar Parallel Arm. *Journal Applied Acoustics*. 66: 1151–1161.

[5] Alam, M. S. and Tokhi, M. O. 2007. Design of Command Shaper Using Gain-Delay Units and Particle Swarm Optimisation Algorithm for Vibration Control of Flexible Systems. *International Journal of Acoustics and Vibration*. 12(3): 99–108.

[6] Khairudin, M., Mohamed, Z., Husain, A. R. and Ahmad, A. 2010. Dynamic Modelling and Characterisation of a Two-Link Flexible Robot Arm. *Journal of Low Frequency Noise, Vibration and Active Control*. 29(3): 207-219.

[7] Tian, Q., Zhang, Y. Q., Chen, L. P. and Yang, J. 2009. Two-Link Flexible Arm Modelling and Tip Trajectory Tracking Based on The Absolute Nodal Coordinate Method. *International Journal of Robotics and Automation*. 24: 103-114.

[8] Olalla, C., Leyva, R., El Aroudi, A., Garces, P. and Queinnec, I. (2010). LMI Robust Control Design for Boost PWM Converter. *IET Power Electronics*. 3(1): 75-85.

[9] Mohamed, Z., Martin, J. M., Tokhi, M. O., Sa da Costa, J. and Botto, M. A. 2005. Vibration Control of a Very Flexible Arm System. *Control Engineering Practice*. 13(3): 267-277.

Performance and exhaust gas emission of gasoline engine fueled by gasoline, acetone and wet methanol blends

Y Prayogi[1] , Syaiful[2], and N Sinaga[3]

[1,2,3] Mechanical Engineering Department Diponegoro University Semarang, Indonesia

E-mail: yogaizze77@gmail.com

Abstract. Increased motor vehicles have resulted in increased fuel consumption and emissions. Pramium fuel has an octane value of 88. Acetone 110 octane value and wet methanol 113 octane value. Mixing of premium acetone and wet methanol is expected to improve engine performance and reduce emissions. Research using gasoline engines with electronic fuel injection (EFI) systems. The results of the testing, there was a 12.62% increase in brake power. Brake specific fuel consumption (BSFC) has increased by 4.35% compared to P100. The increase in BSFC was due to the calorific value of acetone and wet methanol which was lower than the premium. The resulting CO and HC emissions have decreased. The decrease in CO emissions reached 87.5%. The lowest level of HC produced reaches 46 ppm. Mixing acetone and wet methanol into the premium can improve engine performance. It can be seen from the increase in brake power produced. Exhaust emissions are produced better. CO and HC levels have decreased.

1. Introduction

Vehicle growth has a serious impact on a variety of things, including increased emissions and fuel consumption. Vehicle fuel consumption is affected by the acceleration process. Low octane values reduce vehicle acceleration. Fuels that contain low octane values make fuel consumption increase. The higher the octane value of the fuel, the better the engine performance [1].

Acetone and wet methanol are derivatives of alcohols that have high octane values. Ashraf and Chang mix acetone on fuel. The results obtained show an increase in engine performance. Torque and break power on the test machine has increased, while the CO content in emissions has decreased significantly [2][3]. Mixing acetone containing water in gasoline engines can reduce CO and NO_x emissions [4].

Mixing of octane number 88 fuel with 110 octane number methanol on 1000cc EFI 4 cylinder gasoline engine.The results show an increase in engine performance and emission reduction [5]. This oxygen level in methanol influences engine performance [6]. The addition of methanol makes it easy to start the engine when it's cold. The resulting CO and HC emissions have decreased. [7]. Methanol and fuel mixtures on gasoline engines with exhaust gas recirculated (EGR) systems can increase torque and brake power. BSFC has decreased due to the addition of the EGR system [8].

The use of acetone and wet methanol as a premium fuel mixture is expected to improve engine performance and reduce emissions. Engine performance includes the measurement of Brake Power and BSFC. Emissions include measuring CO and HC levels. This research was conducted to test whether the performance and emissions produced using the mixture are better or not.

2. Method

Performance and emission test equipment on gasoline engines are arranged according to figure 1. The fuels used are P (premium), A (acetone) and W (wet methanol). The mixed variable between acetone, wet methanol and premium is in percent (P100), (A10, P90), (A10, W5, P85), (A10, W10, P80), (A10, W15, P75), (A20, P80) , (A20, W5, P75), (A20, W10, P70), (A20, W15, P65), (A30, P70), (A30, W5, P65), (A30, W10, P60), (A30, W15, P55) Supply fuel mixture using the EFI system. Fuel flow is measured to determine the fuel consumption of the engine. Engine output is connected to the DYNOmite Land & Sea dynamometer type water brake to measure engine brake power. Measuring exhaust emissions using Stargass 898 gas analyzer. Measurement of exhaust emissions is used to determine the levels of HC and CO.

Table 1. Fuel properties

Characteristics	Wet Methanol	Gasoline	Acetone
Octane number	113.34	88.80	110
Moisture Content (% v)	24.88	0.003	-
Viscosity	0.46	0.22	0,39
Mass density (kg / m3)	793.2	774	802,6
Calorific Value (MJ / kg)	21.73	42.69	20,89
Flash Point (° C)	10.7	-7.22	18
Latent heat (KJ / Kg)	1139	328.91	519

Table 2. Engine specifications

Engine Toyota Kijang 7 K	
Engine type	Gasoline
Production	Toyota
Number of Cylinders	4
Engine capacity	1798 cc
Number of valve	(SOHC) 8 valves
Maximum power	94 hp - 5000 rpm
Maximum torque	155 Nm – 3200 rpm
Fuel system	EFI

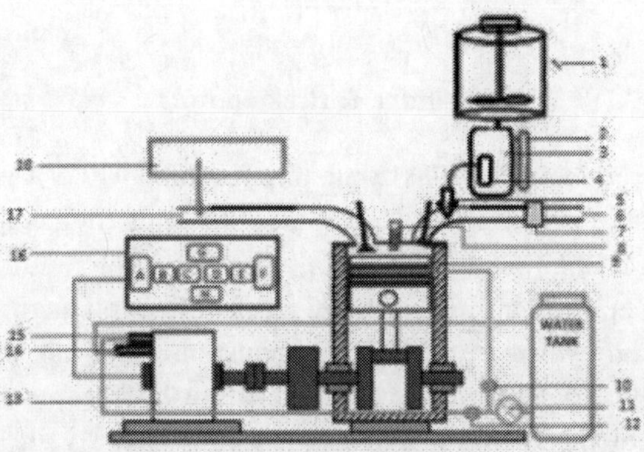

Figure1. Experimental *set up*

1. Fuel Mixer
2. Burret
3. Fuel tank
4. Fuel pump
5. Injector
6. Intake manifol
7. Throtel
8. Spark Plug
9. Cylinder block
10. Cooling water valve
11. Water pump
12. Load valve
13. Dynamometer
14. Outlet water
15. Inlet water
16. Display panel
17. Exhaust
18. Gas analyzer

Testers were carried out on 2500, 3000, 3500 and 4000 rpm engine speed variants with premium fuel mixture, acetone and wet methanol. The test engine uses the Toyota Kijang 5k engine, 4 stroke with 4 cylinder capacity of 1798 cc. Fuel flow rate is measured using a stopwatch to determine the volume of fuel being burned. Fuel mixing is done using a mixer (1). The fuel that has been mixed homogeneously is then flowed to the tank (3). In the tank there is a burette (2) and a fuel pump (4). Buret is used to determine the volume of fuel used. When the gasoline engine works, loading is carried out with a constant load. Measurement of torque and power engine using DYNOmite Land & Sea type water brake dynamometer (13). All measurement data is informed by the display panel (16). In the intake manifold and exhaust manifold the thermocouple is installed to measure the air temperature. Gas analyzer stargass 898 (18) is installed in the exhaust manifold to measure HC, CO, CO_2 and O_2 emissions.

3. Results And Discussion

3.1. Analysis of Machine Performance Testing

The test results in Figure 2, show an increase in brake power as engine speed increases.

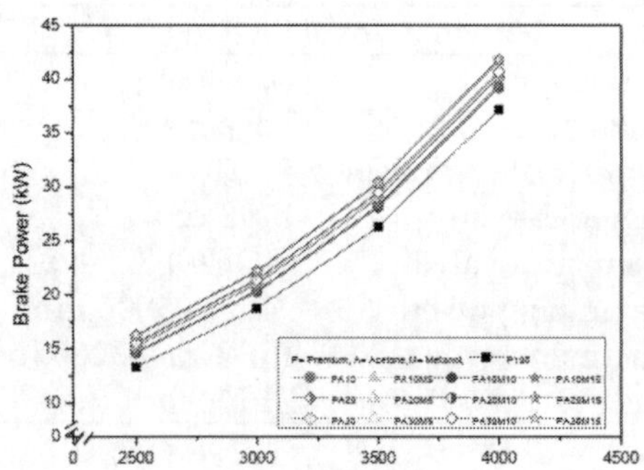

Figure 2. Brake power

The highest increase occurred in the PA30M15 mixture reaching 41.85 KW at 4000 rpm engine speed. The premium produces Brake power of 37.16 KW at 4000 rpm engine speed. The increase in brake power that occurred reached 12.62% with the use of a mixture of premium, acetone and wet methanol. Variation of acetone and wet methanol fuel mixture has an effect on engine power increase. Methanol has a low ratio of carbon and hydrogen. The high combustion speed and the high LHV cause increase the engine power output[9]. Oxygen levels and octane values affect the brake power produced by the engine. Acetone and wet methanol have high oxygen and octane values. Mixing acetone and wet methanol into the fuel causes better combustion and increases Brake power [10]. The water content in wet methanol accelerates the rate of combustion. Addition of acetone increases the peak pressure of combustion results [11].

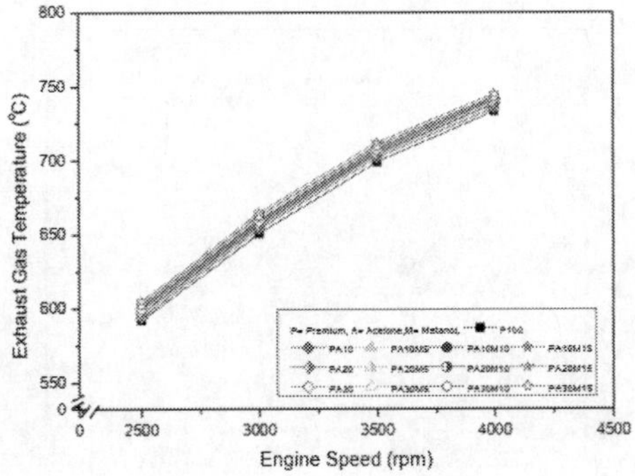

Figure 3. Exhaust gas temperature

Test results on figure 3. There is an increase in temperature in the engine exhaust gas. Increasing the temperature of the exhaust gas occurs in each mixture. The highest increase in exhaust gas temperature occurs in the PA30M15 mixture. The increase in exhaust gas temperature reaches 1.63%. PA30M15 mixture produces exhaust gas temperature of 745 °C at 4000 rpm engine speed. Premium produces exhaust gas temperature of 733 °C at 4000 rpm engine speed. Increased exhaust gas temperature is influenced by high latent heat. Latent heat of acetone and wet methanol is higher compared to premium. Acetone and wet methanol have a higher combustion rate than premium, this causes the combustion of fuel in the cylinder to be more perfect. Oxygen levels in acetone and wet methanol also affect the heat produced in the combustion process [10].

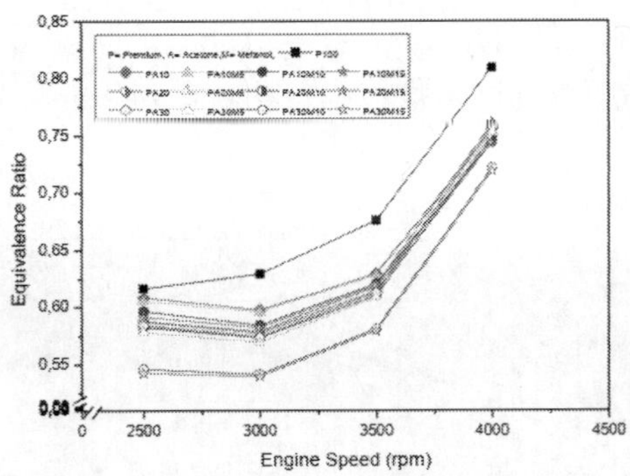

Figure 4. Equivalence ratio

The fuel entering the combustion chamber must be gas. Fuel is more flammable in gas than liquid. The easier the combustible fuel mixture produces maximum energy efficiency [12]. The combustion reaction occurs because there is fuel and oxygen. Fuel and oxygen must be mixed homogeneously. The more homogeneous the mixture of fuel and oxygen the more perfect the combustion process. Comparison of the mixture of air and fuel is strongly influenced by the type of fuel. The mixture of premium, acetone and wet methanol has increased equivalence ratio at each engine speed. The heat value and the smaller stoichiometric ratio of air-fuel causes the equivalence ratio to increase [9]. Figure 4 shows the decrease in the equivalence ratio in each mixture. Premium shows a result of 0.61 at 2500 rpm engine speed. Penuruan occurs gradually along with the large volume of premium fuel mixtures, acetone and wet methanol. The lowest decay occurred in the PA30M15 mixture at a 2500 rpm engine speed showing a result of 0.54. The decrease reached 11.47%.

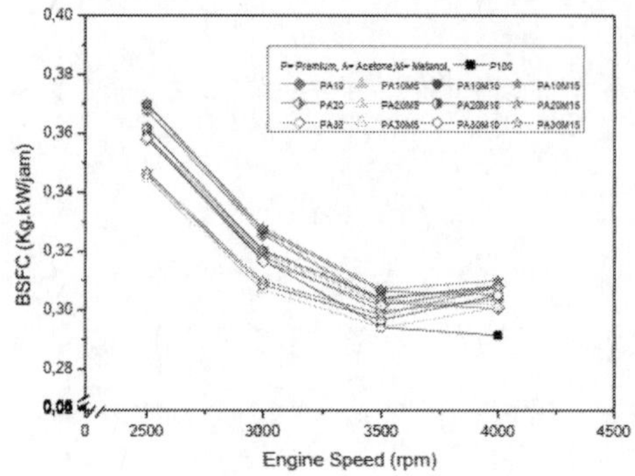

Figure 5. Brake specific fuel consumption

In Figure 5, is the result of BSFC testing. BSFC changes as the engine turns. The lowest BSFC is produced from the PA30M5 mixture at 2500 rpm engine speed. PA30M5 reaches 0.344 KG.KW/Jam at 2500 rpm engine speed. P100 is 0.35 KG.KW/ Hours at 2500 rpm engine speed. P100 experienced the lowest decrease of 0.29 KG.KW/ Hours at 4000 rpm engine speed compared to the others. The PA10M15 mixture produces the highest BSFC compared to other mixtures in each engine rotation. The average increase in PA10M15 mixed BSFC reached 4.35% compared to P100. This is due to the calorific value which is lower than gasoline.

Low heat values do not contribute to heat energy during combustion in the cylinder [11]. The low energy content of the wet methanol, acetone and gasoline mixture causes an increase in BSFC. Energy content and heat value of ethanol / methanol which is lower than gasoline causes an increase in fuel consumption compared to using gasoline [13]. A mixture of acetone butanol ethanol (ABE) on gasoline engines improves engine performance and BSFC. BSFC has increased because ABE has a lower low heating value (LHV) [14].

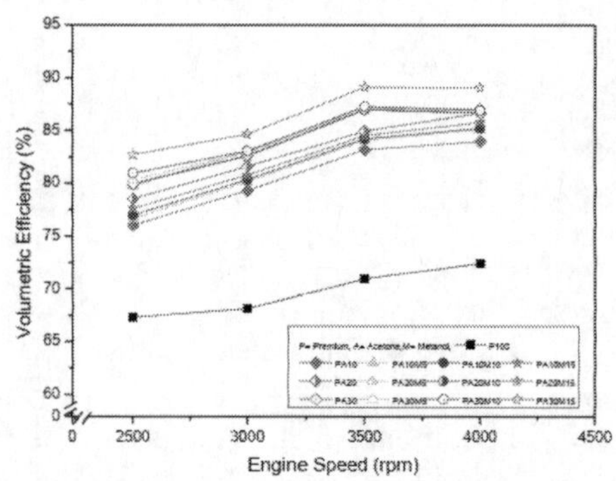

Figure 6. Volumetric efficiency

The test results show an increase in volumetric efficiency in each mixture. The biggest increase occurred in the PA30W15 mixture. These results are due to Acetone and wet methanol having higher evaporation heat than premium. The cooling effect of evaporation of acetone and wet methanol causes the temperature at the intake manifold to be low and increases volumetric efficiency [15]. stated that methanol has a higher combustion temperature than gasoline. More evaporation heat than gasoline causes the intake manifold temperature to be lower and increases volumetric efficiency. [16].

3.2. Data and Analysis of Exhaust Gas Emissions

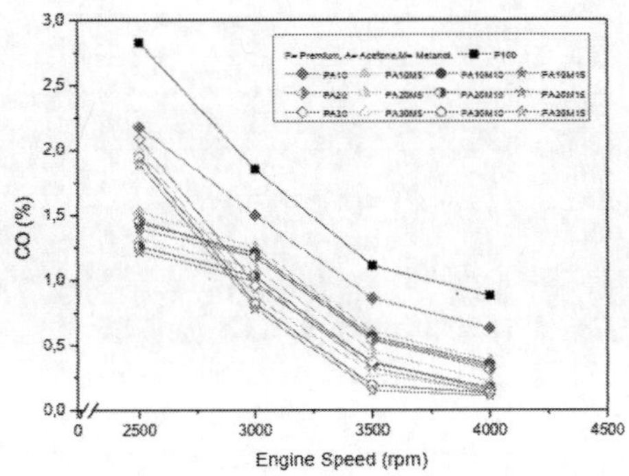

Figure 7. Emission CO

Figure 7 shows the CO level in the exhaust gas decreases in each engine rotation. Highest reduction of CO levels occur in PA30M15 mixture. PA30M15 CO levels reached 0.11% at 4000 rpm engine speed. The drop was influenced by the combustion process and the oxygen content of the wet methanol and acetone [10]. The water content in wet methanol can reduce CO reactions in the combustion process [17].

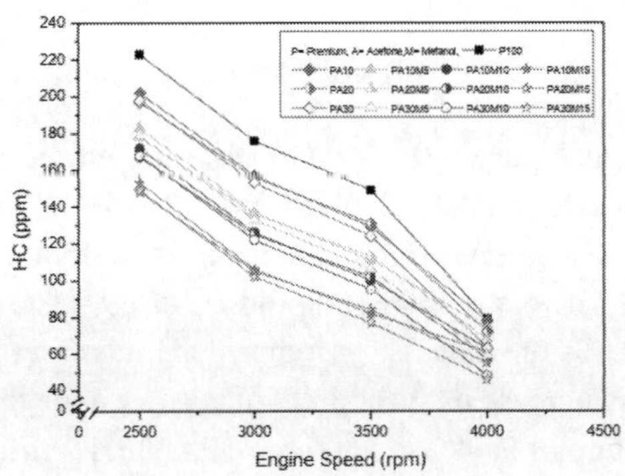

Figure 8. HC Emission

Figure 8, shows that HC emissions formed from combustion products have decreased in each mixture. In the use of P100 HC fuel which is formed reaches 79 ppm at 4000 rpm engine speed. The most significant decrease in PA30M15 mixture reached 46 ppm at 4000 rpm engine speed. The decline reached 41.77%. The decrease in HC that occurs due to more oxygen levels in the combustion reaction. Acetone and wet methanol contain oxygen compounds. When acetone and wet methanol are mixed at a premium. There is an increase in oxygen levels in the combustion chamber. Increased oxygen levels trigger more hydrogen and oxygen reactions than hydrogen with carbon [18].

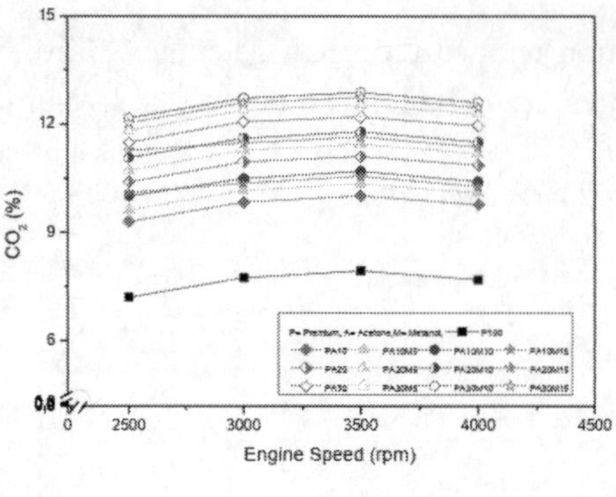

Figure 9. CO_2

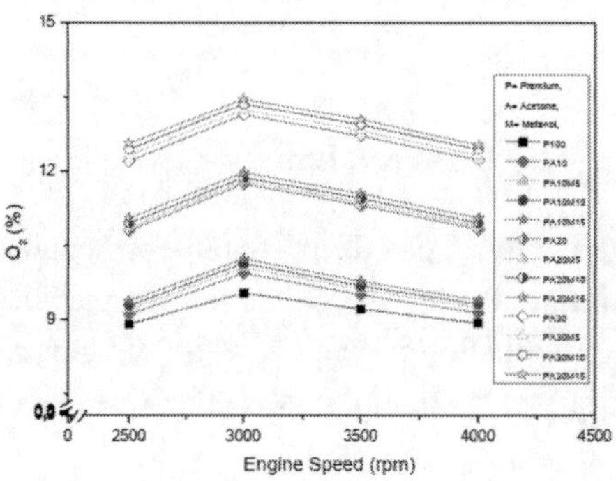

Figure 10. O_2

CO_2 emission results increase. In Figure 9, PA30M10 experienced the highest increase compared to other mixes. CO_2 levels produced by PA30M10 reached 12.88% at 3500 rpm engine speed. P100 mixture CO_2 levels reached 7.93% at 3500 rpm engine speed. At 4000 rpm engine speed, CO_2 levels decrease. PA30M10 produces 12.6% CO_2 and P100 produces 7.69% CO_2.

In Figure 10, the O_2 content in the exhaust gas emissions increases in each mixed variable. The increase occurred along with the increase in the amount of acetone and wet methanol. the least O_2 levels were produced in the P100 mixture of 9.54% at 3000 rpm. PA30M15 produced 13.46% O_2 at 3000 rpm engine speed.

Increased O_2 levels and CO_2 levels are influenced by oxygen compounds on alcohol derivatives. Addition of acetone and methanol increases combustion efficiency. Combustion efficiency increases brake thermal efficiency (BTE) and CO_2 emissions increase. CO emissions levels decrease [19]. Alcohol fuel causes rising CO_2 emissions. The ratio of oxygen in alcohol produces combustion efficiency and causes higher CO_2 emissions [20]. Acetone and methanol solutions contain several percent oxygen. When the solution undergoes a combustion process the residual results from combustion will result in an increase in O_2 and CO_2 levels [10].

4 Conclusion

The main conclusions obtained in the research on the performance and emissions of exhaust gas from EFI gasoline engines with a mixture of premium acetone and wet methanol fuels: [1] Gasoline engine performance has increased with the addition of acetone and wet methanol mixtures into premium. Acetone and wet methanol have higher octane values compared to premium. The increase in octane value in mixed fuels results in increased brake power produced. [5] The exhaust gas emission produced on the test machine has decreased. It can be seen from the decreasing levels of CO and HC contained in the combustion residual gas. O_2 and CO_2 levels have increased. Oxygen levels in wet methanol and acetone reduce the levels of HC and CO formed and increase levels of O_2 and CO_2.

5 References

[1] Anderson, J. E., DiCicco, D. M., Ginder, J. M., Kramer, U., Leone, T. G., Raney-Pablo, H. E., & Wallington, T. J. (2012). *High octane number ethanol–gasoline blends: Quantifying the potential benefits in the United States. Fuel,* **97,** 585.

[2] Elfasakhany, A. (2016). *Performance and emissions analysis on using acetone–gasoline fuel blends in spark-ignition engine. Engineering Science and Technology, an International Journal,* ***19*(3)**, 1224.

[3] Chang, Y. C., Lee, W. J., Lin, S. L., & Wang, L. C. (2013). *Green energy: water-containing acetone–butanol–ethanol diesel blends fueled in diesel engines. Applied Energy*, **109**, 182.

[4] Li, Y., Nithyanandan, K., Meng, X., Lee, T. H., Li, Y., Lee, C. F., & Ning, Z. (2017). Experimental study on combustion and emission performance of a spark-ignition engine fueled with water containing acetone-gasoline blends. Fuel, 210, 133–144. doi:10.1016/j.fuel.2017.07.058

[5] Syaiful, M. N., & Bae, M. W. (2006). *Effect of Hot EGR On Performence And Exhaust Gas Emissions of EFI Gasoline Engine Fueled By Gasoline And Wet Methanol Fuel Blends.*

[6] Elfasakhany, A. (2015). *Investigations on the effects of ethanol–methanol–gasoline blends in a spark-ignition engine: performance and emissions analysis. Engineering Science and Technology, an International Journal*, **18(4)**, 713.

[7] Liu, S., Clemente, E.R.C., Hu, T. and Wei, Y., 2007. *Study of spark ignition engine fueled with methanol/gasoline fuel blends. Applied Thermal Engineering*, **27(11-12)**, pp.1904.

[8] Rohadi, H., Syaiful and Bae, M.W., 2016. *Effect of cooled EGR on performance and exhaust gas emissions in EFI spark ignition engine fueled by gasoline and wet methanol blends*. In AIP Conference Proceedings (Vol. **1737**, No. 1, p. 060005). AIP Publishing.

[9] Iliev, S., 2015. *A comparison of ethanol and methanol blending with gasoline using a 1-D engine model. Procedia Engineering,* ***100***, pp.1013.

[10] Elfasakhany, A. (2016). *Investigations on performance and pollutant emissions of spark-ignition engines fueled with n-butanol–, isobutanol–, ethanol–, methanol–, and acetone–gasoline blends: A comparative study. Renewable and Sustainable Energy Reviews.*

[11] Li, Y., Nithyanandan, K., Meng, X., Lee, T.H., Li, Y., Chia-fon, F.L. and Ning, Z., 2017. Experimental study on combustion and emission performance of a spark-ignition engine fueled with water containing acetone-gasoline blends. *Fuel,* ***210,*** pp.133-144.

[12] Wu, H., Nithyanandan, K., Zhou, N., Lee, T. H., Chia-fon, F. L., & Zhang, C. (2015). *Impacts of acetone on the spray combustion of Acetone–Butanol–Ethanol (ABE)-Diesel blends under low ambient temperature. Fuel,* **142**, 109.

[13] Siwale, L., Kristóf, L., Bereczky, A., Mbarawa, M. and Kolesnikov, A., 2014. *Performance, combustion and emission characteristics of n-butanol additive in methanol–gasoline blend fired in a naturally-aspirated spark ignition engine. Fuel Processing Technology,* **118**, pp.318.

[14] Nithyanandan, K., Zhang, J., Li, Y., Wu, H., Lee, T. H., Lin, Y., & Lee, C. F. (2016). *Improved SI engine efficiency using Acetone–Butanol–Ethanol* (ABE). *Fuel,* **174,** 333.

[15] Celik, M.B., Özdalyan, B. and Alkan, F., 2011. *The use of pure methanol as fuel at high compression ratio in a single cylinder gasoline engine. Fuel,* ***90*(4)**, pp.1591.

[16] Eyidogan, M., Ozsezen, A.N., Canakci, M. and Turkcan, A., *2010. Impact of alcohol–gasoline fuel blends on the performance and combustion characteristics of an SI engine. Fuel,* **89(10)**, *pp.2713.*

[17] Olberding, J., Beyerlein, D.C.S., Steciak, J. and Cherry, M., 2005. *Dynamometer testing of an ethanol-water fueled transit van.* SAE Technical Paper (No. 2005-01-3706).

[18] Li, Y., Nithyanandan, K., Lee, T. H., Donahue, R. M., Lin, Y., Lee, C. F., & Liao, S. (2016). *Effect of water-containing acetone–butanol–ethanol gasoline blends on combustion, performance, and emissions characteristics of a spark-ignition engine. Energy Conversion and Management*, **117**, 21.

[19] Li, L., Ge, Y., Wang, M., Li, J., Peng, Z., Song, Y., & Zhang, L. (2015). *Effect of gasoline/methanol blends on motorcycle emissions: Exhaust and evaporative emissions. Atmospheric Environment*, **102**, 79.

[20] Balki, M.K., Sayin, C. and Canakci, M., 2014. *The effect of different alcohol fuels on the performance, emission and combustion characteristics of a gasoline engine. Fuel*, **115**, 901.

IOP Conf. Series: Materials Science and Engineering **535** (2019) 012014 doi:10.1088/1757-899X/535/1/012014

Basic communication skill drill in microteching context to improve the teching skills of civil engineering and planning education students, faculty of engineering, Yogyakarta State University

Suparman[1], Slamet[2], and GNIP Pratama[3]

[1,2,3]Universitas Negeri Yogyakarta, Yogyakarta, Indonesia

E-mail: suparman55@uny.ac.id

Abstract. This study aims to investigate whether the drill method (verbal and non-verbal communication skill drill) is able to improve the basic teaching skills in microteaching context. This is classroom action research. The population consisted of students majoring in Civil Engineering and Planning Education Department, Faculty of Engineering, Yogyakarta State University who were taking microteaching practice course. The sample of the study was class A1 consisted of 11 students. The study was conducted in three cycles. Cycle 1 contained basic verbal communication, cycle 2 encompassed basic non-verbal communication, and cycle 3 covered both basic verbal and non-verbal communication types. Observation and skills assessment sheets were employed. The qualitative descriptive and descriptive statistics methods were then used to analyze the data.The results show that: (1) basic verbal communication skill drill in microteaching practices improves student teachers' basic microteaching skills and (2) basic non-verbal communication skill drill in microteaching practices improve student teachers' basic microteaching skills.

1. Intoduction

One of the essential skills for teacher is communication skills. Communication is the foundation for an effective teaching and learning process [1]. The micro-learning in this study is divided into two stages. The first stage is microteaching practices in which the students are trained on one learning section at one time such as the opening section, closing section, or whilst teaching activity. The second stage is an integrated microteaching practice involving all microteaching skills such as opening classes, delivering materials, interacting with students, using classroom language, administering learning evaluation, and closing the classes. The integrated microteaching practices are carried out in accordance with the YSU microteaching guidelines. It was observed that some student teachers do not have adequate communication skills when they carried out microteachings. To solve this problem, continuous verbal and non-verbal communication skills trainings are necessary. The two communication types are essential to support teaching skills. Microteachings are typically conducted in the form of repeated practices such as practicing how to open and close classrooms as well as to conduct whilst teaching activities. These typical practices are considered ineffective because of the many skills that should be repeated. In the drill method, student teachers are trained on micro-skills such as how to speak clearly and loudly and how to move around the classroom. If these micro-skills are mastered by the student teachers, then it will be easier for them to develop other teaching skills.

1.1 Drill Method

There are several attempts made by experts to define the drill method. Nana Sudjana defines the drill method as a repetition of the same activity in an earnest manner to strengthen certain associations or to perfect and internalize certain skills [1]. The distinctive characteristic of this method is the repetition of the same activity. Further, Toto Hernawo states that the drill method, also called the training method, is a beneficial teaching method to instill and maintain habits, and to develop dexterity, accuracy, opportunities, and skills [2]. In line with the previous opinions, Sugiyanto defines the drill method as a method in which students perform movements according to what the teacher instructs and do it repeatedly [3]. The repetition of this movement is intended to enable movement automation. All definitions of the drill method proposed by the aforementioned experts have one similar characteristic, i.e. repeated training to acquire and internalize certain skills.

1.2 Basic Communication Skills

Communication is the process of transferring an understanding in the form of ideas and information from one to another. It means that communication is more than just transferring meaning but also make the meaning understandable. Communication can be verbal or non-verbal. The former type is the communication that uses verbal media or spoken words in conveying messages. On the other hand, the latter is the type that uses non-verbal media in conveying message such as one's face, eyes, body, touch, sound, space, time, physical appeal, clothing, and environment [4]. Verbal communication is considered effective when meeting the following requirements: (1) building positive environment to develop fun communication, (2) encouraging students to speak up, (3) seeking opinions, (4) sympathizing with students' feelings, (5) reaching agreement, (6) building sufficient eye contact, (7) bonding, (8) leaving a good impression, (9) motivating, (10) showing lively voice and body language, and (11) addressing students with positive names. Meanwhile, non-verbal communication has the following characteristic: (1) continuous in nature, (2) rich in terms of meaning, (3) sometimes confusing, (4) conveying emotion, (5) controlled by norms, and (6) related to culture.

1.3 Microteaching

According to the YSU microteaching guidelines, microteaching is also called internship practice II. The basic competencies in micro teaching are as follows: (1) understanding the basics of microteaching/ Internship II, (2) designing lesson plans, (3) practicing the basic skills of microteaching, (4) practicing the basic skills of the integrated teaching, (5) evaluating microteaching practices. Further, microteaching is carried out in each faculty's study program under the supervision of in-charged lecturers and under the coordination of a study program's and faculty's microteaching coordinator. Microteaching is for the sixth semester students. There are ten components of basic teaching skills. The following components are required to open the lessons: (1) drawing students' attention, (2) motivating students, (3) providing references, and (4) providing apperception. The component of closing the lesson includes: (1) reviewing the material, (2) evaluating student learning outcomes, (3) making conclusions or summaries of the material, (4) providing significant (appropriate, meaningful, and useful) assignments.

2. Method

This is classroom action research. This study sought to improve the student teachers' (S1/ undergraduate program) basic microteaching skills by drilling basic communication skills in microteaching practices. The student teachers would participate in integrated field practices (*PLT*) in vocational schools. The population of this study consisted of students majoring in Civil Engineering and Planning Education, Faculty of Engineering, Yogyakarta State University who were taking microteaching courses in the 7th semester in 2016. The sample of the study was class A2 comprising 11 students who were under the supervision of the research team's members.

This study follows the action research model proposed by Kurt Lewin consisting of four steps: (1) planning, (2) acting, (3) observing, and (4) reflecting. The purpose of the planning step is to prepare all the instruments needed for the second step which involves designing microteaching syllabus, lesson plans, and observation sheets. The plan formulated in the first step was then carried out in the action step. In this step, a microteaching practice was conducted to examine the student teachers' prior communication skills. Afterwards, a series of treatments in the form of drill method was given. The microteaching practice was continued after the student teachers were drilled on the basic communication skills. During the microteaching process, observation was conducted. The observation results of the microteaching prior and after the drills were then compared to examine the changes. The student teachers' skills were also assessed qualitatively. The reflection step involved discussions between the research team members and student teachers on how to improve the microteaching quality. Observation and skills assessment sheets were used to collect the data. The collected data were then analyzed statistically and qualitatively.

3. Result amd Discussion

This part describes the process, results, and discussion of the following aspects: (a) the student teachers' prior microteaching skills, (b) cycle 1 on the basic verbal communication drill, (c) cycle 2 on the basic verbal communication drill, and (d) cycle 3 on the basic microteaching skills practices.

3.1 The Student Teachers' Prior Microteaching Skills

This study starts with the very basic teaching skills, i.e. opening and closing the lessons. This activity is intended to find out the student teachers' prior micoteaching skills. A total of 11 students were given a chance to conduct a 10 minute microteaching. The observation results show that alot of students are not adequate in terms of verbal and non-verbal communication skills. Further, not all components of opening and closing the lessons were included when the students conducted their turns.

Cycle 1 focuses on the basic verbal communication skill drill. The student teachers are trained on how to use verbal communication skills for a ten-minute teaching. The skills trained in this cycle are voice volume, speech flexibility, clarity, and fluency. In a meeting, each students get 2 to 3 turns to practice all those basic skills. The assessment results of Cycle 1 are presented in Table 1.

Table 1. The assessment of the basic verbal communication skill drill

Student Number	The basic verbal communication skill score				Total Score	Mean Score
	Voice Volume	Flexibility	Clarity	Fluency		
1	3	2	2	2	9	2.3
2	3	3	3	3	12	3.0
3	3	3	3	2	11	2.8
4	2	3	3	3	11	2.8
5	4	3	3	4	14	3.5
6	3	2	3	2	10	2.5
7	3	3	2	2	10	2.5
8	3	3	3	2	11	2.8
9	3	3	2	3	11	2.8
10	3	2	3	3	11	2.8
11	4	3	3	4	14	3.5

Total	34	30	30	31	
Mean Score	3.1	2.7	2.7	2.8	2.8

The research team reflects and discusses what steps are needed to improve the microteaching practices based on the observation and assessment. The reflection results suggest that the student teachers should seriously act as teachers or students. In the next cycle, the student teachers are asked to write and memorize a script containing what they will say in the next meeting. The results of this reflection are used to design the next exercises.

Cycle 2 encompasses the basic non-verbal communication skill drill. Based on the Cycle 1 reflection, the student teachers are asked to write and memorize what they will say for the next meeting. The lecturers consistently remind the students to seriously immerse themshelves in their roles either as teachers or students. This cycle focuses on how to give positive responses in non-verbal ways through facial expression, hand movements, human contact, and teacher movement in the classroom. The participants are trained on those skills repeteadly (drill system). The cycle 2 assessment are presented in the following table.

Table 2. The assessment of the basic non-verbal communication skill drill

Student Number	The basic verbal communication skill score			Total Score	Mean Score
	Movement	Flexibility	Attractiveness		
1	2	2	2	6	2,0
2	3	3	3	9	3,0
3	3	3	2	8	2,7
4	3	3	3	9	3,0
5	3	3	2	8	2,7
6	2	3	2	7	2,3
7	3	2	2	7	2,3
8	3	3	2	8	2,7
9	3	2	3	8	2,7
10	2	3	3	8	2,7
11	3	3	2	8	2,7
Total	30	30	26		7,8
Mean Score	2,7	2,7	2,4	7,8	2,6

In cycle 2, the student teachers show serious efforts in taking their roles both as teachers and students. However, their non-verbal communication skills are not yet sufficient. In early performances, communication breakdown often occurs or in other words the stimulus and response are incompatible. For example, when 'the teacher' shows positive/pleased expression, 'the students' does not reciprocate the same sentiment. Therefore, 'the students' are expected to provide stimulus that supports the non-verbal communication development.

Cycle 3 is carried out in the second half of the program. The microteaching practice in this cycle is intended to examine the effect of communication skill drill on the student teachers' performances. The first meeting focuses on training the student teachers specifically on whilst teaching activities or providing explanation of the subject matters. Before conducting microteachings, the student teachers are asked to write down a scenario on how they will ochestrate their classes. The lecturers observe and compare the written scenario and what unfolds in the field. If there is any discrepancy, the lecturers will point the errors out (substantial or technical errors) to the student teachers.

The Cycle 3 observation shows that the student teachers are quite flexible and fluent in teaching. However, the way they explain the subject matters is monotonous. In the second meeting of Cycle 3, the students seems inexperienced in using various teaching media and technology. However, their teaching performances are considered good. They need to practice more on operating LCD projector and designing media presentation.

The third meeting focuses on the student teachers' skills at leading classroom discussions. They are considered insufficient in leading discussion activities. There are often long pauses and even complete stops in the middle of the discussions because 'the teachers' do not know how to coerce their 'students' to participate. In these conditions, the lecturer gives guidance on how to lead the discussions. The student teachers' performances in leading discussions are gradually improved. The observation results show an improvement in verbal and non-verbal communication skills.

4. Conclusion

(1) Basic verbal communication skill drill is considered able to improve the student teachers' basic microteaching skills; (2) Basic non-verbal communication skill drill is considered able to improve the student teachers' basic microteaching skills.

Acknowledgments
In regard with the effectiveness of verbal and non-verbal communication skills to improve microteaching skills, microteaching process should be conducted in three levels: (1) basic communication skills, (2) basic microteaching skills, and (3) basic integrated teaching skills. The drill method is recommended in microteaching courses. The skills in which the students are lacked is practiced repeatedly.

5. References

[1] Muhammad Surya. (2013). *Psikologi guru, konsep dan aplikasinya*, dari guru, untuk guru. Bandung: Alfabeta
[2] Toto Hernowo.Drill and Practice. *Internet: http://totohernawo. blog.m3-access. com/ posts/35929_ DrillAnd-Practice.html*
[3] Sugiyanto (2016). Metode Drill. Internet: http://www.gudangteori.xyz/2016/01 /pengertian-metode-*drill*-latihan-menurut. html
[4] Muhammad Budyatna dan Leila Mona Ganiem. (2012) *Teori Komunikasi antar pribadi.* Cetakan ke-2. Jakarta: Kencana Perdana Media Group

IOP Conf. Series: Materials Science and Engineering **535** (2019) 012015 doi:10.1088/1757-899X/535/1/012015

An analysis of changes in flange surface roughness after being used to tighten a corrugated metal gasket

D Nurhadiyanto[1], S Haruyama[2], Mujiyono[3], and Sutopo[4]

[1,3,4]Universitas Negeri Yogyakarta, Yogyakarta, Indonesia
[2]Yamaguchi University, Japan

E-mail: didiknur@uny.ac.id

Abstract. A corrugated metal gasket is not reusable once the plastic deformation has occured. Meanwhile, flange might be reusable if it has not corroded. This research aimed to find out the surface roughness and contour shape of the flange after being used to tighten the gasket. This study employed a method of measuring the surface roughness before and after the first, second, third, and fourth tightening. The flange used had a surface roughness of Ra 3.5μm, Ra 2.5 μm and Ra 1.5 μm. The measurement of surface roughness was based on the standard of JISB0601-2001. The findings show that no significant change occured on the flange surface roughness. However, the corrugated contour shape did occur in the surface roughness after use.

1. Introduction

The use of gasket materials has been terminated following a ban on producing and using asbestos worldwide. A replacement for asbestos materials, therefore, is necessary. Asbestos are used in gasket to prevent leakage in pipes. Gasket is placed between two flanges connected to pipes. To date, the development of gasket materials from metal, i.e. the corrugated metal gasket, has taken place. The function of the corrugated gasket is to decrease the clamping force and produce a spring effect to create a sealing line on the flange.

Research on corrugated metal gasket has been carried out by many scholars. Saeed et al. (2008) initiated the development of the corrugated metal gasket. The gasket had a spring effect and produced a high local stress contact that created a sealing line on the flange [1]. Afterwards, Haruyama et al. (2009) studied the minimum contact width on a gasket using simulation [2]. They found the most optimum contact width in their study. The contact width can be used as the design parameter of the corrugated metal gasket. Through an experiment, it was found that the leakage can be minimized by increasing the contact width. Nurhadiyanto et al. (2012) searched for the most optimum dimension of metal gasket to prevent leakage using a simulation analysis of Finite Element Method (FEM) [3]. Then, the simulation and experiment results were compared. The comparison showed that both results matched each other. With regards to the experiment, the helium leak test indicated that the leakage was not found. Haruyama et al. (2013) studied the force per unit length, contact width, and contact stress for gasket in contact with the flanges with different surface roughness by using FEM [4]. The result of FEM analysis was compared with that of the experiment. The softer the flange surface roughness, the smaller the leakage created using helium leak test will be. The contact between a rough flange surface and a very soft gasket resulted in the rough gasket. Nurhadiyanto et al. (2018) studied the corrugated metal gasket coated with copper [5]. The coating was aimed at softening the gasket surface without changing its stiffness so that the contact stress remains the same. The result of the simulation was proven with the experiment which showed that the gasket performance increased. The leakage in the copper-coated gasket decreased compared to that in typical gaskets.

These previous studies indicate that the flange surface roughness influences the leakage in the connection. Gasket is easily scratched by the rough flange surface because gasket materials are softer than those of flanges. The gasket surface will become rough as it is in contact with the flange surface. As a consequence, gasket materials fill the roughness of flange surface and repair the leakage. After that, the gasket will deform because of the clamping force. A gasket can only be used for one tightening process. However, the characteristic of flange remains unknown. Is there any change in the flange surface roughness? Is there any change in the flange surface contour? This study aimed at answering these questions.

2. Method

The gasket used in this study was a corrugated metal gasket with two crests and two troughs. Figure 1 shows the corrugated metal gasket coated with copper. The corrugated shape of the gasket was produced using cold press molding. The gasket dimension used was JISB2404 (2006) which was 25A in size and 1.45mm in thickness [6]. Figure 1 shows the gasket used in this study. The gasket material used was SUS304 which was tested using JISZ2241 (2011) [7]. The test result showed the nominal stress of 398.83 MPa, the elastic modulus (E) of 210 GPa and the tangent modulus of 1900.53 MPa.

Figure 1. The Corrugated Metal Gasket.

The flange was based on JISB2220 (2004) with 10K pressure and 25A diameter [8]. Figure 2 shows the flange used in this study. To prevent distortion, the lower and the joint parts of the flange were carefully welded. The detail information on the welding can be found in Haruyama et al. (2013) [9]. The types of flange were differentiated based on the surface roughness parameter (Ra): 1.5μm, 2.5μm, and 3.5μm.

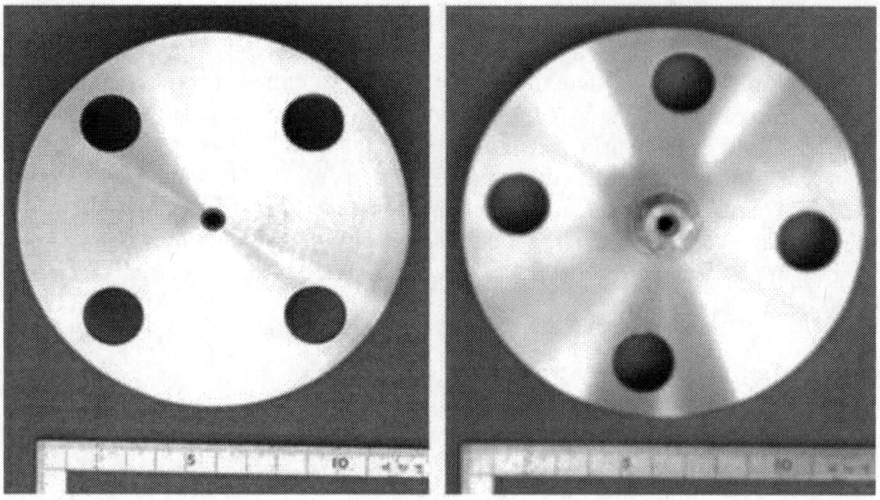

Figure 2. Flange

Previous studies have reported that the leakage was likely to occur as the surface roughness value increased. The simulation result showed that the smaller roughness value increased the contact width [4]. A VH-Z250 series microscope was used to measure the roughness of the gasket surface before and after the tightening process [10]. The microscope helped the researchers measure the grooves of the surface roughness after tightened by the flange. Figure 3 shows the illustration of the grooves of the surface roughness (Haruyama et al., 2014) [11].

As shown in Figure 3, some grooves on the gasket were found after the tightening between the gasket and the flange. The tightening used bolts which resulted in the high local contact stress on the gaskets crest and trough. Equation (1) shows that the total number of grooves width which mean a real contact width between flange and gasket as explained by Haruyama et al. (2014) [11].

$$\mathrm{CW} = \sum_{i=1}^{n} h_i \tag{1}$$

The tools for measuring the surface roughness of the flange are shown in Figure 4. The JISB0601-2001 standard used to measurement of the surface roughness [10]. The data of roughness were measured at four points of contact width which were opposite to one another. The final data were in the form of the average surface roughness of those four points. Further explanations on the measurement of the surface roughness can be seen in Haruyama et al (2013).

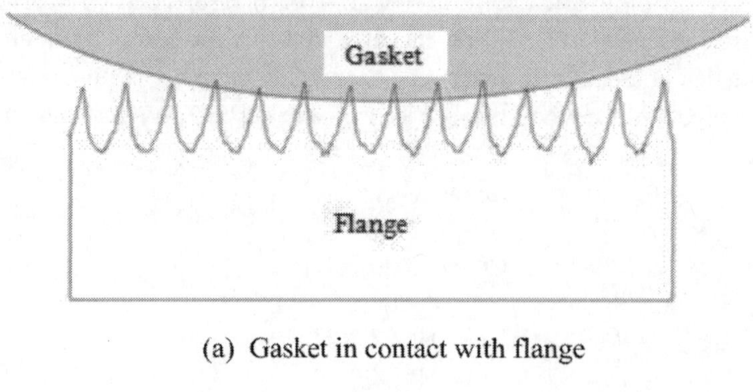

(a) Gasket in contact with flange

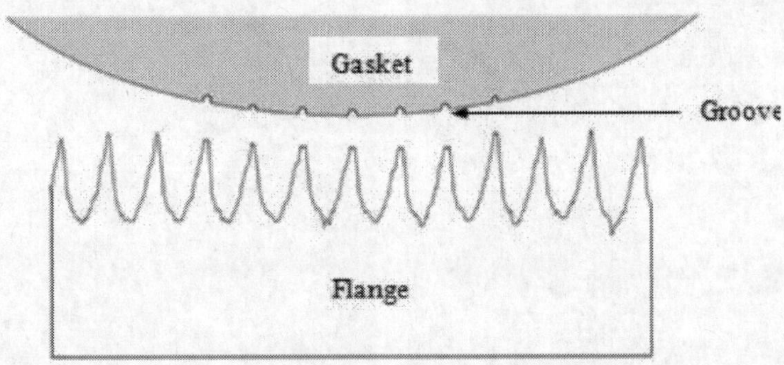

(b) Grove formed after contact

Figure 3. Gasket in contact with flange and groove formed after contact (taken from [11])

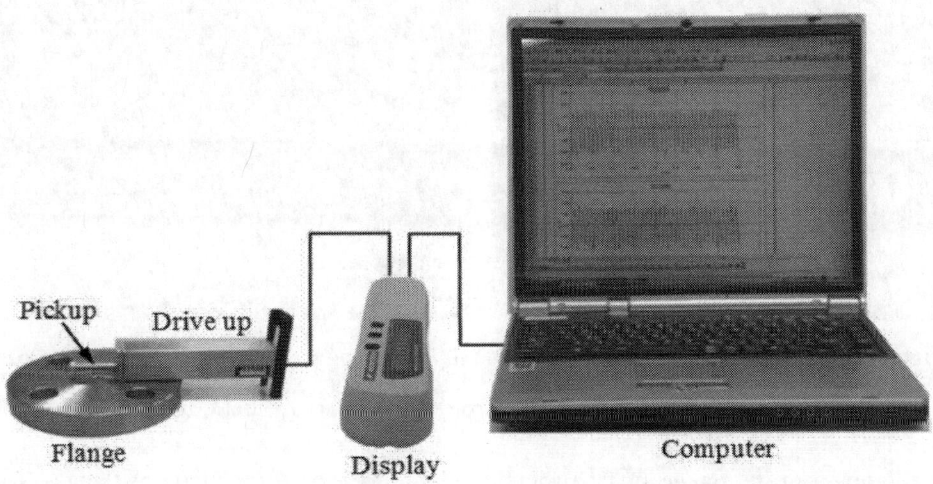

Figure 4. The measurement of the surface roughness (taken from [4])

There are three kinds of flanges which have the surface roughness of Ra 3.5μm, Ra 2.5μm, and Ra 1.5μm. The measurement of the flange surface roughness was carried out in three situations: before and after the flange use for the first, second, third, and fourth tightening. This study shows the changes the surface roughness and the changes the shape of the surface contour.

3. Results and Discussion

Gasket was deemed not having any surface roughness before use. Once the gasket was used, the grooves appeared because of the flange surface roughness. Figure 5 shows the gaskets surface roughness before and after it was used for tightening. The grooves which appeared showed that gasket materials filled up the flange roughness which eventually repaired the leakage because it was covered by the gasket. It was the purpose of coating the outer surface of the gasket with the softer materials.

(a) before tightening (b) tightening with the flange of Ra1.5 μm

(c) tightening with the flange of Ra 2.5 μm (d) tightening with the flange of Ra 3.5 μm

Figure 5. The surface roughness of the gasket

The measurement of the flange surface roughness was done on three kinds of flanges: Ra 3.5μm, Ra 2.5μm and Ra 1.5μm. The measurement result of the surface roughness of the Ra 1.5μm flange is presented in the form of image in Figure 5. Meanwhile the statistical result is presented in Figure 6, which shows the changes in contour.

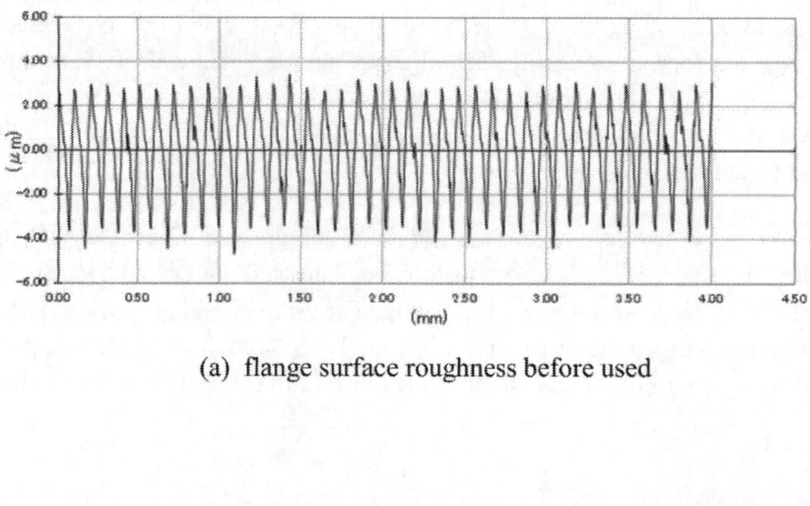

(a) flange surface roughness before used

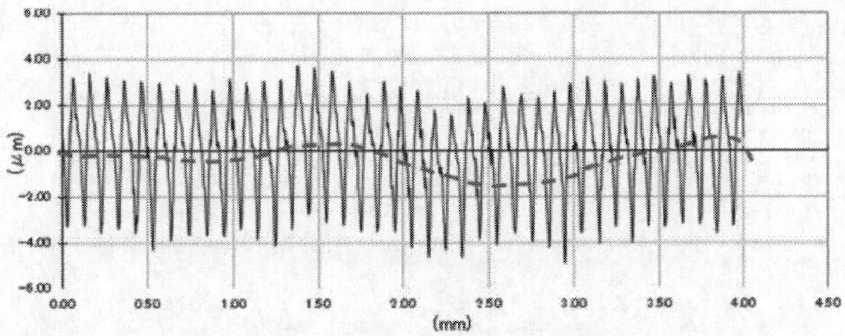

(b) flange surface roughness after the first tightening

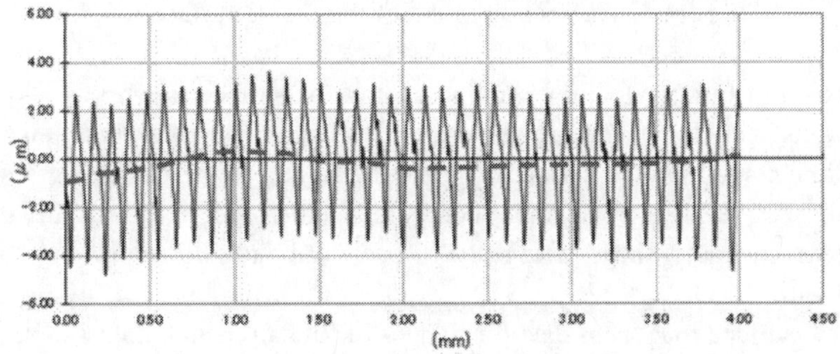

(c) flange surface roughness after the second tightening

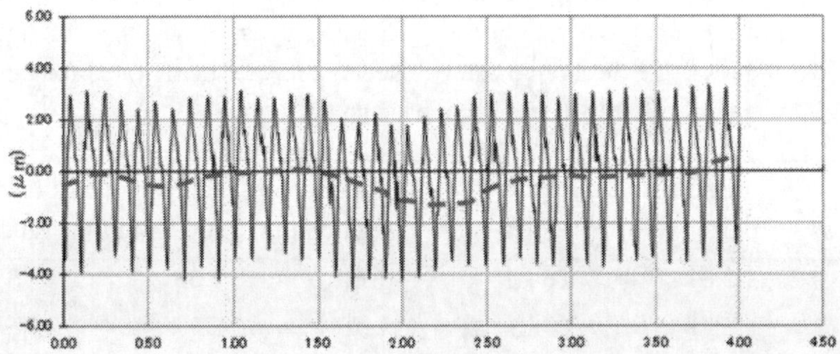

(d) flange surface roughness after the third tightening

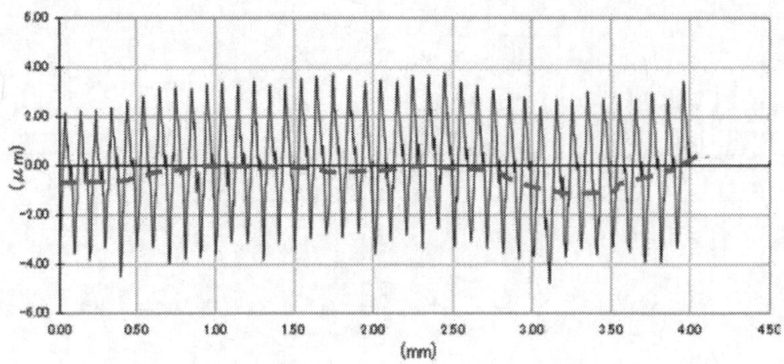

(e) flange surface roughness after the fourth tightening

Figure 6. The measurement of the flange surface roughness

Initially, as shown in Figure 6(a), the midline contour of surface roughness was in the form of a stright line. It means that the surface roughness was symmetrical: the crests and troughs were in balance. The midline contour of the surface roughness changed as the flange was used for tightening as shown by the dashed red line. After the second, third and fourth treatment, the surface roughness experienced contour changes. These changes, however, did not show how big the flange surface roughness value was.

In the same way, surface roughness data for flanges of Ra 2.5μm and 3.5μm were obtained. Surface roughness data for Ra 2.5μm and 3.5μm showed the same trend as that for Ra 1.5 μm. Statistical data of experimental results show that there is a very small change in the flanges surface roughness. The average surface roughness data for flanges can be seen in Table 1. This data is the average data from four points. It is seen that the surface roughness before and after being used for tightening does not change too much. This shows that the flange can be used again after being used for tightening.

Table 1. The flange surface roughness

Surface Roughness	Before used	First tightening	Second tightening	Third tightening	Fourth tightening
Ra 1.5	1.61	1.60	1.60	1.58	1.58
Ra 2.5	2.78	2.78	2,76	2.77	2.76
Ra 3.5	3.91	3.91	3,86	3,75	3.80

4. Conclusions

Based on the above discussion, the following conclusions can be drawn; (1) Gaskets will change the degree of the surface roughness after being used and after tightening process. This surface roughness due to contact with high pressure causes a softer material gasket to fill the surface roughness of the flange; (2) The value of flange surface roughness does not change much before and after use. Flanges can be reused if there is a gasket replacement as long as there is no corrosion; (3) The contour of the flange surface roughness changes. It is no longer in a straight line, but the surface roughness does not change.

IOP Conf. Series: Materials Science and Engineering **535** (2019) 012015 doi:10.1088/1757-899X/535/1/012015

5. References

[1] Saeed, H.A, Izumi, S., Sakai, S., Haruyama, S., Nagawa, M., Noda, H. 2008. Development of New Metallic Gasket and its Optimum Design for Leakage Performance vol 2 *Journal of Solid Mechanics and Material Engineering* p 105

[2] Haruyama, S., Choiron, M.A., Kaminishi, K., A. 2009. Study of Design Standard and Performance Evaluation on New Metallic Gasket, *Proceeding of the 2nd International Symposium on Digital Manufacturing, Wuhan China*, p 107

[3] Nurhadiyanto, D., Choiron, M.A., Haruyama, S., Kaminishi, K. 2012. Optimization of New 25A-size Metal Gasket Design Based on Contact Width Considering Forming and Contact Stress vol 6 *International Journal of Mechanical and Aerospace Engineering* p 343

[4] Haruyama, S., Nurhadiyanto, D., Choiron, M.A., Kaminishi, K. 2013. Surface Roughness of flange Contact to the 25A-size Metal Gasket by Using FEM Simulation vol 307 *International Journal of World Academy of Science, Engineering and* Technology p 665

[5] Nurhadiyanto, D., Mujiyono, Sutopo, and Ristadi, F.A. 2018. Simulation Analisis of 25A-Size Corrugated Metal Gasket Coated Copper to Increase Its Performance vol 307 *IOP Conference Series: Materials Science and Engineering* p 1

[6] Japanese Standards Assosiations. 2006. *Dimensions of gaskets for use with the pipe flanges.* JISB2404

[7] Japanese Standards Assosiations. 2011. *Metallic materials-Tensile testing-Method of test at room Temperature.* JISB2241

[8] Japanese Standards Assosiations. 2004. *Steel Pipe Flanges.* JISB2220.

[9] Haruyama, S., Nurhadiyanto, D., Choiron, M.A., Kaminishi, K. 2013. Influence of Surface Roughness on Leakage of New Metal Gasket vol 84 *International Journal of Preesure Vessels and Piping* p 349

[10] Handysurf E-35A/B User Manual, 2

[11] Haruyama, S, Nurhadiyanto, D., Kaminishi, K. 2014. Contact Width Analysis of Corrugated Metal Gasket Based on Surface Roughness vol 856 *Advanced Materials Research* p 92

IOP Conf. Series: Materials Science and Engineering **535** (2019) 012016 doi:10.1088/1757-899X/535/1/012016

Driving control module for low cost industrial automated guided vehicle

ES Maarif[1] and T Moyo[2]

[1,2]Automation Laboratory, Astra Manufacturing Polytechnics, Jakarta, Indonesia

E-mail: eka.samsul.maarif@gmail.com

Abstract. Automatic Guided Vehicles (AGVs) are computer-controlled "driverless" mobile vehicles equipped with optical, magnetic or laser guidance systems for automated functionality, categorized as either load carrying or load towing. AGV consists of several parts, such as the frame for pulling or lifting weights, movement balancer wheels, battery housing, collision avoidance sensors, driving module and others. But, the most important part of AGV is the driving module which is a module that functions as a movement guide, consisting of path detection sensors, driving motors and controllers. This Paper presents low cost a driving control module for low cost AGV that used in automotive manufacture industry. The AGV uses Raspberry Pi instead of PLC as controller, infrared sensor to detect line and industrial standard brushless DC (BLDC) motor as main actuator. PID control applied in the movement of the BLDC motor speed so that the AGV can follow the path smoothly and precisely. As a result, the AGV can follow the path with speed 30 m/min on a straight road and 30 m/m at a turn with a deviation 5 mm off the track.

1. Introduction

Material handling problem in industrial field consists of how to transport raw materials, partially manufactured products and goods between different locations of manufacturing systems, warehouses, etc. Depending on the kind of products to handle and the transport to perform, there are different solutions in the market. For instance, we can mention belt, roller and vertical conveyors, elevators, material handling robots and automated guided vehicles (AGVs) [1].

AGVs also have several advantages inherent to their design, such as the reduction of product damage from removal of human error, the ability to travel into hazardous areas without concern for operator safety, the ability to automatically track and record product movement, the reduction of labour, and the flexibility and adaptability especially prevalent in laser guided systems. AGV also plays main rule in Flexible manufacturing systems (FMS), which are equipped with several CNC machines and AGV-based material handling system are designed and implemented to gain the flexibility and efficiency of production [2].

The basic principle of the operation of traditional AGV system is in a pre-defined route placed on the floor or located in the floor, which is followed by reading device, which is a part of each traditional AGV. In addition, other support devices are moving toward the basic functionality "tracking pre-defined route "to a higher level. They are safety bumpers, optical scanning devices, obstacle sensors, audio and visual warning devices, Wi-Fi connection, GPS modules, remote controls, communication via smartphone, camera systems and others [3]. There are several ways of navigation and guidance technologies of AGVs, which range from physical guide-paths using laser, magnetic

IOP Conf. Series: Materials Science and Engineering **535** (2019) 012016 doi:10.1088/1757-899X/535/1/012016

tape, optical sensors, wire, to gyroscope based inertial guidance and wireless. The latter provides the advantage that can be easily modified [4].

AGV generally has PLC or microcontroller as the main control device. Suman Kumar [5], at his paper Increase in Efficiency using PID Control of an Automated Guided Vehicle for Product Ware House, has develop microcontroller based AGV for his thesis. He uses ATmega16 to read line sensors that produce analogue signals and the microcontroller controls the speed of DC motors with L298 motor diver. He also applies PID control to adjust the motor rotation speed [5]. Martin Kajan [6], at his paper "Control of Automated Guided Vehicle with PLC SIMATIC ET200S CPU", has been researching the use of PLC SIMATIC ET200S as AGV controller. The AGV communicates with the human machine interface (HMI) via industrial wireless LAN with the use of PROFINET and PROFISAFE communication standards. Direct connection of distributed control systems through industrial Ethernet can be easily implemented via PROFINET [6].

This paper focuses on developing a driving module at a low price but can still be used in industrial environments, especially in automotive manufacture industries and has a drive that is strong enough to pull loads up to 500 kg. This AGV uses a mini PC as an alternative to expensive PLC prices. Raspberry Pi is chosen, as it has flexibility in programming languages, sufficient input output, able to be communicated with the touch screen panel and allows further development of wireless communication.

2. Method

2.1 Design

2.1.1 Driving Design. A general AGV system essentially consists of vehicle peripheral on site component as well as stationary control system. The main components of AGV system are [7]: (1) Vehicle; (2) Guidance path system; (3) Controller.

The driving unit module in this paper covers the discussion of vehicle and guidance path systems. In the sub-section vehicle describes the AGV work system as a puller or load carrier. This will have an impact on the discussion of the method of controlling the direction of the AGV movement, AGV's speed and the motor drive that is suitable for the job. In the sub-section guidance path describes the path sensors system used on the driving module and motor control to be able to follow a predetermined path. The last sub-section discusses the controller and the motor control method on AGV's driving module.

The load that must be moved by AGV in the automotive manufacturing industry is relatively heavy, with a consumption of up to 500 kg. Thus, it is more likely to design AGV as a drag load rather than as a heap load. Because by giving a heap load it requires more powerful motor and greater power consumption. Figure 1 shows the frame design of AGV.

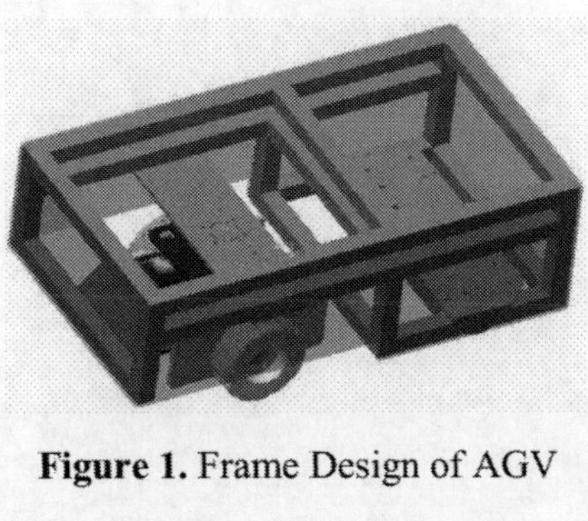

Figure 1. Frame Design of AGV

When designing the vehicle, there are four basic ways of providing drive and steering [8]. These are show in Figure 2 below. Some types are easier to build, others have better steering characteristics. Two of these designs are fully symmetrical and may be operated in both directions.

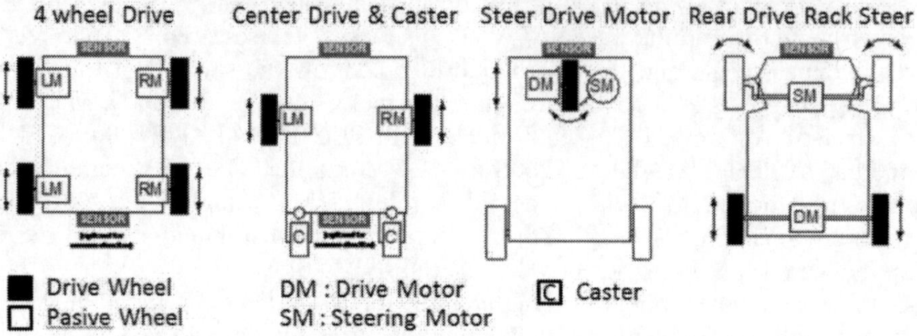

Figure 2. Chassis Design

Table 1 gives list the characteristics of each design.

Table 1. Chassis Design Characteristics

Design	Simplicity	Steering	Reversible
4 Wheel Drive	Simplest	Coarse	Yes
Centre Drive & Caster	Simple	Precise	Yes
Steer Drive Motor	Medium	Very Precise	No
Rear Drive Rack Steer	Complex	Precise	Difficult

In this project center drive was chosen because of the ease of control and accuracy of movement. The prime mover of AGV can be selected between using a DC motor or brushless DC (BLDC). DC motors can be controlled with PWM drivers and signals [6][10], but DC motors does not have enough torque to load up to 500 kg. BLDC motors are high torque that the most commonly used motors in the industrial world, such as automotive, medical automation and instrumentation equipment. The BLDC motor does not use brush (brush) for magnetic field changes (commutation), but done electronically commutated.

Driving module system of the AGV movement which is designed in the drive unit sub assembly as in Figure 3 and the parts of the sub unit driving module unit and its function explained in Table 2.

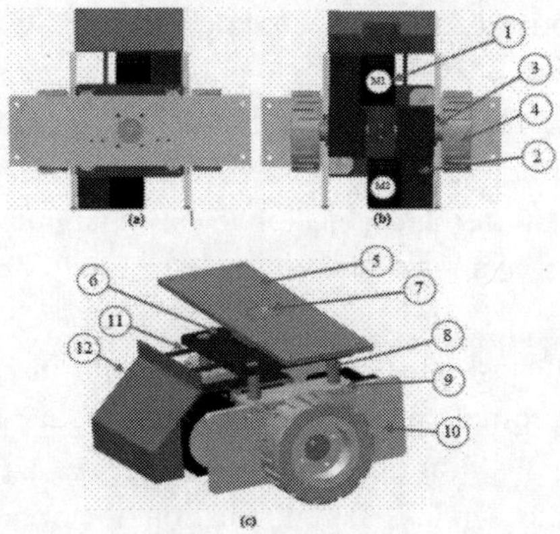

Figure 3. Drive Unit AGV (a) Top view, (b) Down view, (c) Isometric view

Straight movements or turn movements is regulated by the rotation speed of each motor. For straight movements, the motor speed of the right and left must be the same, while for the movement turns right and left motor speed is distinguished. For example, if the AGV will turn right then the motor speed is set so that the right motor is slower than the left motor so that the driver's mechanical movement will turn right, and vice versa.

Table 2. Components of Driving Module

No	Component	Function
1	Brushless Motor	Actuator for wheel (2500 rpm)
2	Gearhead Motor	Speed reducer and torque multiplier
3	Shaft wheel	Motor and wheel connector
4	Wheel	AGV mover
5	Base 3	Driving unit to Chassis connector
6	Base 2	Stabilizer base for right and left wheel
7	Shaft 2	Base 2 and 3 joint
8	Shaft 1	Joint unit and base 2 connector
9	Joint Unit	Base connector
10	Base gearhead	Gearhead attachment
11	Spacer	Base 2 to Sensor's bracket attachment
12	Sensor's Bracket	Sensor attachment

2.1.2 Path Guidance. Selecting the guidance path system is based on the floor conditions that AGV will go through and how often the route changes will be taken. Fixed path (floor wire, magnetic tape, reflective tape, colour line) guidance is beneficial when the system path is well defined, simple and permanent. Open path (laser, inertia) guidance provides significant benefit with regard to path flexibility. Some AGV uses Radio-frequency identification (*RFID*) which monitors the AGV systems report amount, location, and identify of inventory located on the AGV. It can also report the location of products in real time [11].

This driving module uses line lines as paths because it is the cheapest and easiest to determine which sensors are used. Therefore, a sensor is needed that can detect color differences. Color sensor usually contains an infrared transmitter and detector in one integrated components, they are infrared (transmitter) and photodiode (receiver).

2.1.3 Controller. PLC as controller which is commonly used in industry, offers system stability and ease of use [12]. However, to get complete features like analogue input, High Speed Input readings (to read encoder), analogue outputs, PWM or pulse output, the prices will be very high. Moreover, if the AGV system will be developed into wireless communication, the price will be more expensive.

In this project, Raspberry Pi mini PC was chosen as an alternative to PLC because it has a cheaper price and complete features including analogue input and output, PWM output, communication to the touch screen panel to built-in WIFI. Raspberry also offers the ease of selecting programming languages such as C, C ++, or Python. The language used in Raspberry Pi is the Python programming language and the help of several studies that have been carried out by several people who have uploaded to social media and Google. Figure 4 shows control block diagram of the system.

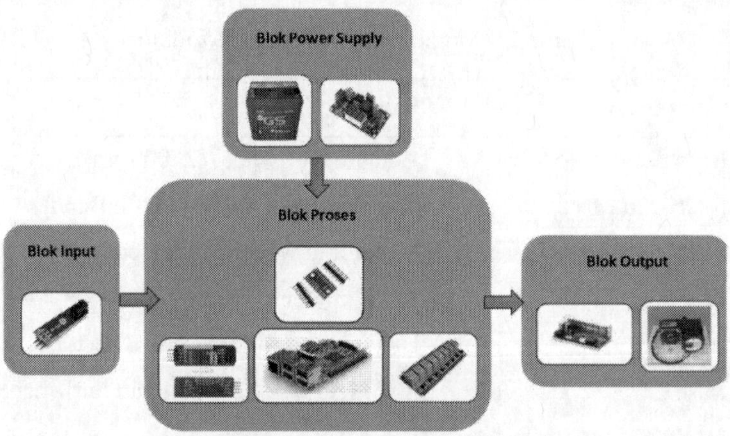

Figure 4. System Block Diagram

Block Power Supply has battery and power supply module. Block Input has TCRT5000 Sensor as optical sensor to detect color line. Block Control has Biderectional Voltage, Raspberry Pi and Relay module for switching. Block Output has BLDC motors and the drivers

2.2 Development

2.2.1 Speed of the AGV. The AGV criteria that will be made includes a number of things, such as 70 I/O including backup I/O, can be connected to touch panel and can carry a pull load of 500kg. The allowed speed standard on standard safety in automotive industry shown at Figure 5.

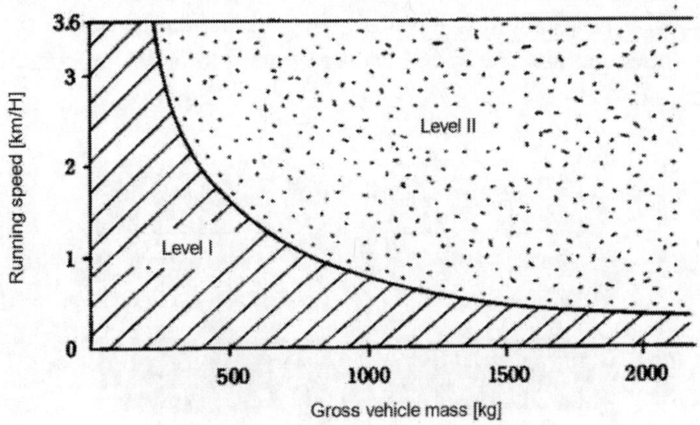

Figure 5. Maximum Speed Chart

In the picture above, to draw a load of 500kg, the maximum speed is only 1.8 km / h.

$$1.8 \text{ km / h} = (1.8 \times 1000) \text{ m} / (1 \times 60) \text{ min} = 1800 \text{ m} / 60 \text{ min} = 30 \text{ m / min} \quad (1)$$

Whereas, in the manufacturing industry, the average allowable speed is 30 m / min when on a straight path and only 30% speed at turning path.

With a wheel diameter size of 140 mm, a rotational speed of 2500 Rpm motor and a gear head scale of 30: 1 it can be seen that the maximum speed of the AGV will be made, namely: Wheel = 140 mm, Wheel circumference = 440 mm = 0.44 m; *v* motor = 2500 Rpm; gear head = 30: 1 ratio.

$$\text{v motor / v wheel} = \text{gear head comparison} \quad (2)$$
$$2500 / x = 30/1$$
$$x = 2500/30$$
$$x = 83 \text{ Rpm}$$

So, speed the wheel now after going through the gearhead is 83 Rpm. With the wheel speed that has been known, it can be seen that the maximum speed of the AGV that will be made is:

$$v \text{ AGV} = \text{v wheel x around the wheel} \quad (3)$$
$$= 83 \text{ Rpm} \times 0.44 \text{ m}$$
$$= 36.52 \text{ m / min}$$

From the speed 30m/min, it can be seen that the maximum PWM signal that can be given to the driver is as follows:

$$100 / \text{PWM} = 36.52 / 30 \quad (4)$$
$$\text{PWM} = (30 \times 100) / 36.52$$
$$= 3000 / 36.52$$
$$= 82.14$$

The maximum PWM signal that can be given to each driver is 82.14% of the maximum voltage which is 5volt or equal to 4.1volt.

2.2.2 Electrical Control. In designing the input device consists of 16 TCRT5000 sensor modules which are used to read the AGV movement path. The input device will be connected to the IO expander as the IO unit of Raspberry PI 3, which will then be connected to the pin contained in the process device, namely Raspberry PI 3. The input device read line path and controller execute the program that was created for activate the output device, that is, the brushless motor driver by giving the rotation direction command and the speed used by each motor in the condition or data from the read input device, Figure 6.

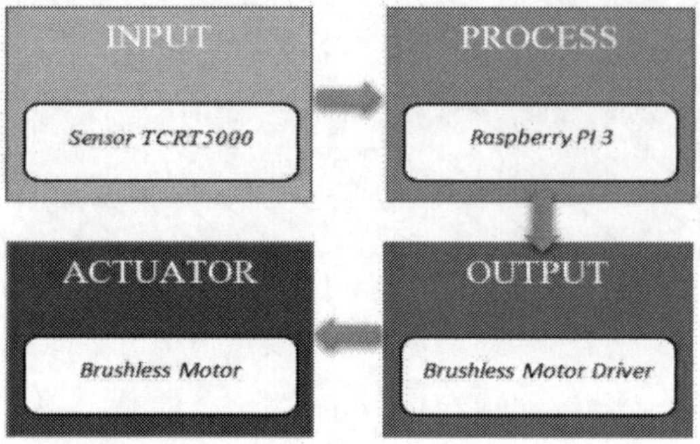

Figure 6. Electrical Controller Design

The TCRT5000 sensor is connected to the expander IO which will then be connected to Raspberry PI 3 through bidirectional voltage first. This is done because the input for Raspberry PI 3 must not exceed 3.3 volt and with the bidirectional voltage module, this original voltage of 5 volts will be changed to 3.3 volt so that it can be used by Raspberry PI 3. Then the output of the relay module will be directly connected to Raspberry PI 3 because the 3.3 volt voltage is enough to activate the relay module that will be used to control the motor driver.

2.2.3 Programming PID Controller. Programs written in Python are typically much shorter than C or C ++ programs, due to several considerations: (1) High-level data types are used to declare complex operations in a single statement; (2) Statement grouping has been completed with indentation instead of starting and ending suffixes; (3) No arguments or variable declarations are needed.

Separate program blocks with an indentation arrangement. So to enter sub-programs in a block, the sub-programs are placed one or more spaces from the column of a program block. A command sentence is an instruction that can be executed by a Python interpreter. Previously there were two types of command sentences, namely print and value declaration. Generally, comments contain information about the usefulness of a function. The syntax is the "#" fence sign.

Variable is a storage value that can be used in a program. Variables can be changed according to instructions made: (1) Variable Name, the programmer uses the variable name in accordance with the description of the contents of the variable and the variable is also a symbol that represents a certain value. Creating variables in Python is very simple; (2) Keywords define the rules and structure of the language, and they cannot be used as variable names.

The stability of AGV movement in running following the line is very important because it affects if the AGV movement is unstable, it will be in terms of security of AGV usage. For this reason, PID control tuning is needed. Tuning this PID control aims to determine the proportional, integrative, and derivative control action parameters of AGV. PID (Proportional – Integral – Derivative controller) controller applied to PWM motor signal. It is a feedback mechanism controller that is usually used in industrial control systems. A PID controller continuously calculates the error value as the difference between the desired set point and the measured process variable. The controller tries to minimize the error value at any time by setting the control variable to a new value determined by the PID formula.

The output of the PID control signal is formulated as below [5]:

$$u(t) = K_P \cdot e(t) + K_I \cdot \int_0^t e(t)dt + K_D \cdot de(t)/dt \quad (5)$$

In a discrete time the control signal output is formulated:

$$PID = Kp \cdot e(k) + Ki \cdot [e(k\text{-}1) + e(k)] \cdot Ts + Kd \cdot [e(k) - e(k\text{-}1)] / Ts \quad (6)$$

whare: PID is Control output signal; Kp is Proportional gain, tuning parameters; Ki is Integral Gain, tuning parameters; Kd is Gain Derivatives, tuning parameters; Ts is sampling time; e (k) is Error (Sp - PV); e (k-1) is Last Error; Sp is Setpoint; PV is Process variable (reading sensor weight).

At this project, tuning done without identify the AGV mechanism or dynamic analysis. Figure 7 is an illustration of the sensor weighting (sensor reading error value) on the AGV to be created:

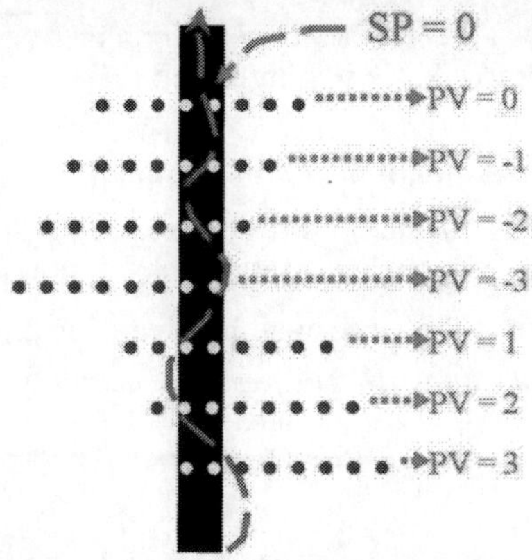

Figure 7. Sensor Weighting Error Value

The ideal condition of AGV occurs when the AGV condition in PV = 0 (eg the condition of the sensor value = 1111110000111111, the value 0 represents the sensor about the line). Or SP = 0 is the ideal condition of AGV. If PV is not equal to 0, it means that the robot is not in an ideal condition and that means there is a signal error. In this error conditions the PID formula will determine the results of the control signal.

PID drive program controls motor speed with reference to sensor reading data on the lines tracking program. It is divided into several function programs, namely: (1) Motor driver control program: This program is used to provide input to the motor driver in controlling the right and left motor rotation directions rotating in CW, CCW, or stop which will be used by the PID program; (2) Sensor weight mapping program: This program is used to map the weight of error values from sensor readings and determine the ideal conditions of AGV. The weight of the sensor reading must be different between sharp turns and not then this weight will be an error condition of the AGV position. The error condition of this program will be processed by the PID formula (1) Switching speed program: This program is used to adjust the speed used by AGV in running so that the AGV speed in the straight line will be different from the path when turning. By using the sensor reading conditions, a switching speed system on AGV can be designed. The following reference paths are used in switching AGV speed programs.

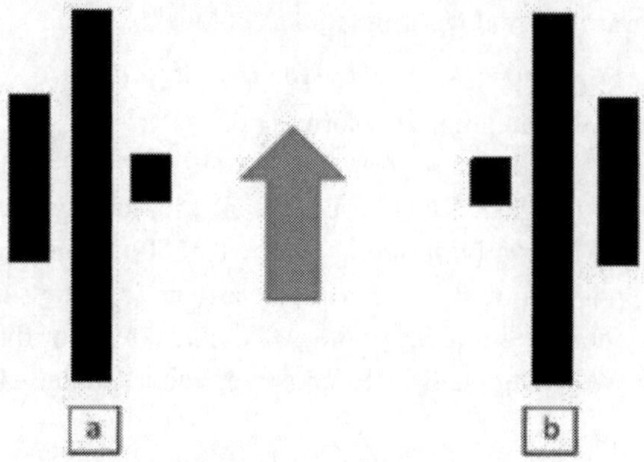

Figure 8. Switching speed reference path

Condition (a) will be used for minimum speed requirements or speed when turning, condition (b) will be used to change AGV speed to maximum speed or speed when on a straight line; (2) PID program: This program is used AGV to stabilize itself in walking following the line that has been made. The PID formula will process the error weight given by the sensor weight mapping program and will determine the results of its own control signal according to the weight of the sensor that is read.

2.3 Testing

2.3.1 Test Line Tracking Program. This test aims to find out the line tracking program that is functioning properly and can provide data that can be used as parameters to give commands to the driving program. There are 2 decimal numbers where the decimal number above is the result of the left sensor reading (SL) and the bottom is the result of conversion from the right sensor reading (SR). The picture below is a sample image of the sensor reading that has been converted to decimal and displayed with the print command on IDLE Python.

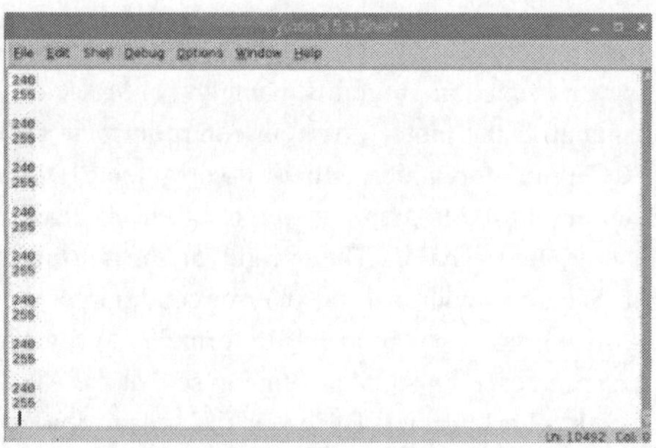

Figure 9. Testing the tracing line program

The picture above is a test image of a tracking line program displayed in Python shell where there is a reading of SL = 240 and SR = 255. These results are decimal values of sensor readings SL =

11110000 (sensors 5, 6, 7 and 8 detect the line) and SR sensor = 11111111 (right sensor does not detect line).

2.3.2 PID Motor Drive Test. The driving program testing aims to find out whether the data from the TCRT5000 sensor that is available can be processed by the driving program and the PID program so that it can run the motor to follow the line as expected. The results of the sensor readings will be processed by the PID program which will determine the results of its own control signal in accordance with the weight of the sensor that is read, then it will control the movement of AGV by giving a PWM signal to each motor driver in running following the available line. In this test, conditions that are read by the TCRT5000 sensor will be displayed in the input column where the red colour is a sensor that is reading the black line and the green colour is a sensor that does not read the black line. Then there is a speed column which is the AGV wheel speed from the control results of each motor driver controlling the speed of the wheel in following the available line. The picture below is the driving test data by showing the actual speed given to each wheel based on the weight of the sensor that is read.

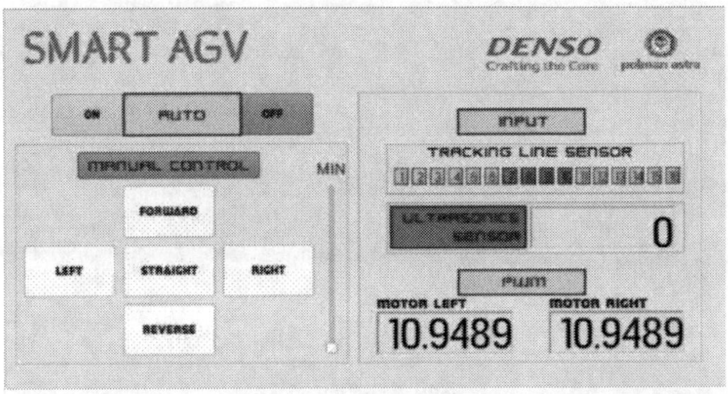

Figure 10. Testing of Motor Drive programs 1

The picture above is a picture of testing the driving program where there are TCRT5000 sensor readings displayed on the AGV HMI which are sensors 7, 8, 9, and 10 detecting the line (AGV means in ideal conditions or PV = 0) and the speed used in each right and left motor driver is equal to 10.9 m / min so that the AGV will run straight along the line. This speed will still be given and does not change before the error condition changes or AGV is not the ideal condition. The following are the results of testing on AGV when it is not in its ideal condition. An error condition that will be used as a test sample is PV = -5 so that the PID calculation and the speed should be given to each wheel as in the following figure.

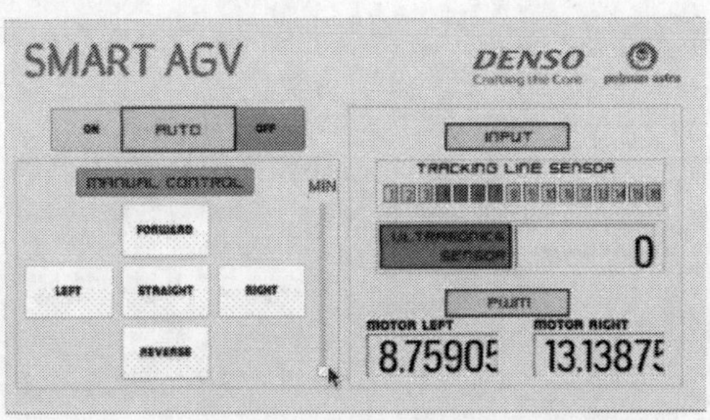

Figure 11. Testing of Motor Drive programs 2

The picture above shows the results of TCRT5000 sensor readings displayed on the AGV HMI namely sensors 4, 5, 6 and 7 detect the line with the sensor weight value used in the condition is -5 and the speed used on the right wheel is 13.1 m / min and the left wheel is 8.7 m / min so the AGV will turn left to get the ideal AGV condition.

3. Result and Discussion

The test results from the previous section explain that the control program of the AGV driving module using Raspberry PI 3 which has been made has indicated the conformity of the actual working principle of AGV with the principles of AGV's work on the design concept that has been made. AGV can walk following the line as expected, using Raspberry PI 3 on a line colour using the TCRT5000 sensor and the selection of replacement components can be used on the next AGV because it has been proven to replace the previous component. Figure 12 shows sensor of the AGV.

Figure 12. Line Sensor

The AGV can run at a speed of 30 m / min on a straight line and 10m / min on a path with a large defection of 1cm. this is influenced by the distance of each sensor used on AGV, which is 1 cm. the following explanation about AGV deflection can be seen in the picture.

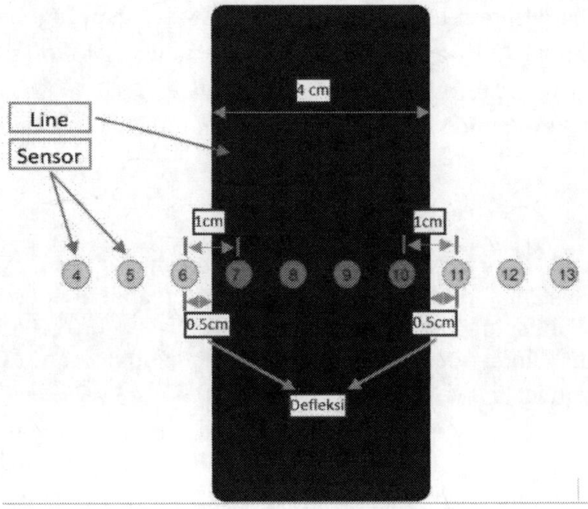

Figure 13. Evaluation of AGV movement deflection

In the picture above it can be seen that the AGV deflection made is affected by the distance between the sensors used. When the AGV runs straight, sensors 7, 8, 9 and 10 will detect the path, but with uneven floor conditions will also affect the movement of AGV in following the line. If the AGV turns left at 0.5 cm it will make the sensor 6 detect the line and the program will correct the AGV condition and will move to the right to get the ideal condition again. Similarly, if the AGV moves to the right and the 11 sensor detects the line, the program will correct the AGV condition and move to the left to get the ideal condition again. So it can be concluded that the AGV movement deflection in following the line is 1 cm seen from the outer distance of the line with the next sensor which is 0.5 cm right deflection and 0.5 left deflection. Figure 14 shows the AGV.

Figure 14. The AGV

4. Conclusion

Control programs for low cost Automated Guided Vehicle (AGV) driving modules using Raspberry Pi 3 with TCRT5000 sensor input and PWM output for BLDC driver as actuators can run as expected.

The stage of making the program using Python 3.4 software, the program is adjusted to the AGV work function, namely: manual AGV activation, AGV automatically follows the line color, and can be adjusted to AGV speed so as to improve the AGV performance.

The AGV speed that is 30 m/min in the straight and 10 in turns with deflection about 5 mm.

5. References

[1] H. Martinez-Barbera, D. Herrero-Perez, "Development of a Flexible AGV for Flexible Manufacturing Systems", Industrial Robot: An International Journal 37/5 (2010) 459–468, Emerald Group Publishing Limited. [DOI 10.1108/01439911011063281]

[2] Y.Seo, P.J. Egbelu, " Integrated Manufacturing Planning for an AGV-Based FMS", Int. J. Production Economics 60Ð61 (1999) 473Ð478

[3] G. Fedorko, S. Honus2, R. Salai," Comparison of the Traditional and Autonomous AGV Systems", MATEC Web of Conferences 134, 00013 (2017) DOI: 10.1051/matecconf/201713400013 LOGI 2017

[4] Llopis-Albert et al, "Designing Efficient Material Handling Systems Via Automated Guided Vehicles (AGVs)", Multidisciplinary Journal for Education, Social and Technological Sciences ISSN: 2341-2593, Vol. 5 Nº 2 (2018): 97-105 | 97, http://polipapers.upv.es/index.php/MUSE/

[5] S.K.Das, "Increase in Efficiency using PID Control of an Automated Guided Vehicle for Product Ware House", Thesis, Department of Mechanical Engineering, National Institute of Technology, Jamshedpur-831014, June 2016

[6] M. Kajan, L. Mrafko, F. Duchoň, P. Hubinský, J. Šovčík, "Control of Automated Guided Vehicle with PLC SIMATIC ET200S CPU", American Journal of Mechanical Engineering, 2013, Vol. 1, No. 7, 343-348 Available online at http://pubs.sciepub.com/ajme/1/7/38

[7] S.K Das, M.K.Pasan, "Design and Methodology of Automated Guided Vehicle-A Review" , IOSR Journal of Mechanical and Civil Engineering (IOSR-JMCE, Special Issue - AETM'16, Page 29

[8] A. Kumar, "Development of An Automated Guided Vehicles In Industrial Environment ", Int. J. Mech. Eng. & Rob. Res. 2014 ISSN 2278 – 0149 www.ijmerr.com, Vol. 3, No. 1, January 2014

[9] H. Dudeja, L. Bagal, N. Zunjur, S.S. Jagadale, "Mechanical Design of an Automated Guided Vehicle (AGV)", International Journal of Research In Aeronautical And Mechanical Engineering, ISSN (Online): 2321-3051 Vol.3 Issue.5, May 2015. Pgs: 32-40

[10] R. Kapolyo, Bing, Z. Gang, "Research and Design of AGV System Application Based on PLC and RFID", International Journal of Science and Research (IJSR)

[11] S.S. Shingare, O. Swami, K. Mahajan, Y. Junawane, "Modification of Material Handling Process Using Automated Guided Vehicle (AGV)", International Research Journal of Engineering and Technology (IRJET) e-ISSN: 2395 -0056 Volume: 04 Issue: 02 | Feb -2017

IOP Conf. Series: Materials Science and Engineering **535** (2019) 012017 doi:10.1088/1757-899X/535/1/012017

The implementation of resource-based learning to improve creativity in developing Yogyakarta batik motif

K Asiatun[1], S Widarwati[2], Sugiyem[3], and AG Bestari[4]

[1,2,3,4] Yogyakarta State University, Yogyakarta, Indonesia

E-mail: asiatun@uny.ac.id

Abstract. This study aims to 1) investigate the implementation of Resource-Based Learning (RBL) in Batiking Course, and 2) identify RBL's impacts on students' creativity in developing the motifs of Yogyakartan batik. The method employed in this study was Classroom Action Research proposed by Kemmis and McTaggart, involving a cycle of planning, observing, and reflecting. This research was carried out in Fashion Design and Hospitality Education Department, Faculty of Engineering, Yogyakarta State University in the 2017/2018 academic year, from January to June 2018, with 36 students from Fashion Design Study Program participating as the research subjects. The instruments were tested for content validity and the items proved properly designed, while the reliability was assessed using Cronbach's alpha, resulting in an alpha value of 0.525, showing that the instrument items could be used. The obtained data were analyzed using descriptive statistics. The results of this study show that 1) the implementation of RBL in Batiking Course has been carried out in 2 cycles, consisting of Introduction, Main Activities, and Closing, and 2) in the Pre-Cycle, Cycle I, and Cycle II phases, the mean scores of the students' creativity are 29 (less creative), 37 (creative), and 47 (very creative) respectively. Therefore, the number of creative students in this implementation has increased from only 28% to 94%.

1. Introduction

The Fashion Design Education Study Program of Yogyakarta State University offers a Batik Technology course with 2 credits of practice to enhance the graduates' competences. Even though this course is a practical one, the teaching and learning process is accompanied by supporting theories, including the basic concept of batik, Indonesian batik motifs, materials for batiking, tools for batiking, and techniques in batiking (*tulis*, *cap*, and *colet*). Theory mastery will determine the students' learning achievement in this Batik Technology course.

In fashion design field, creativity is something that each graduate must possess for their competences are the realization of their ability and creativity. Munandar defines creativity as an ability to elaborate ideas [1]. Batik is a work of art that has been recognized by the world as the cultural heritage of the Indonesian people. Prasetyo asserts that batiking is one of the ways in making clothing materials that refers to a coloring technique using *malam* (wax) [2]. Susanto classifies batiks based on their motifs: geometric and non-geometric [3].

Nowadays, batik is used in not only formal events but also any non-formal occasions. Besides, batik is one of creative industries that need to be developed through innovations, particularly through motif variations that can always meet the market's demand. To do so, the Fashion Design Education students must possess extensive knowledge of batik motifs. This knowledge can be acquired if they use various resources in learning, thus adding inspiration to the making of their work.

However, there are still many students who are reluctant to look for learning resources about batik, particularly about the motifs of Yogyakartan batik, causing their works to be very simple. This condition certainly hinders the learning process. Therefore, this study tries to overcome this problem by implementing Resource-Based Learning using both mass and electronic media as the learning resources.

Beswick in Seel [4] affirms that Resource-Based Learning (RBL) involves students' active participation with the use of varied resources (books, journals, newspapers, multimedia, websites, societies, people) to eventually motivate them to learn by trying to find as much information as possible. RBL is not the only method implemented at schools for there are some other methods available. RBL is only an example of those methods so that there is no need to eliminate the other methods. Nasution mentions the characteristics of RBL as follows: (1) RBL makes use of any information resources as learning resources, including audio-visuals, and gives chances to the students to plan their learning activities by taking into account the available resources [5]. By this description, it does not mean that lecturing method can be eliminated; instead it can be combined with other methods to reach a certain purpose; (2) RBL displays a wide variety of learning resource, including resources from societies, human environments, museums, organizations, printed materials, libraries, audio-visual tools, etc. The students must be taught the techniques for doing field work and making use of libraries and reference books, so that they can be more confident in learning; (3) RBL aims to change the students' passiveness in learning traditionally into active learning by paying more attention to interests and involvement in the learning process. For this purpose, the materials taught should be more meaningful and varied; (4) Unlike any other conventional methods that require the students to learn monotonously, RBL supports the use of varied materials, methods, and communication medium that eventually will increase the students' learning motivation; (4) RBL lets the students to learn on their own pace and capability; (5) RBL offers more flexibility in terms of the time and space for learning; (6) RBL encourages the development of students' confidence in learning that is expected to be able to support a lifelong education.

In all this time, the students' works are assessed only based on the final product of their works so that the expected competences cannot be achieved. In this study, the students' works are evaluated through a portfolio assessment beginning from the very first time they start their works until they finish it. This method allows direct correction when there is a mistake made by the students in any stages of the learning process. The implementation of portfolio assessment can intensify the collaboration and communication between a student and the other student, a student and his/her teacher, a teacher and the other teacher, and between the portfolio developer and the teacher. Through the implementation of RBL and portfolio assessment, it is expected that the students' insight into batik will be wider so that their creativity in developing batik motifs can be stimulated and the desired competence can be achieved. The purposes of this study are to implement Resource-Based Learning in Batiking Course and to improve the Fashion Design Education students' creativity in developing the motifs of Yogyakartan batik.

2. Method

This study was a classroom action research, referring to the Kemmis and McTaggart's model that includes a cycle of planning, observing, and reflecting. The subjects of this study were 36 students of Fashion Design Education Study Program enrolling in Batik Technology Course in the 2017/2018 academic year.

The Planning phase was the initial mechanism focused on the preparation of everything needed in this study, both the things related to the subject matter and the research design with its technical administration. The second phase, Observing, was the phase in which RBL was implemented and its implementation was observed, and the students' learning activities were assessed. The last phase, Reflecting, was an evaluation phase intended to assess the learning achievement, viewed from both the process and the final product perspectives.

2.1 Research Instruments

The instrument used to obtain data in this study was test. The test administered in this study was a psychomotor test that assessed the students' creativity in designing Yogyakartan batik motif. Table 1 presents the items assessed to measure the students' creativity in designing batik motifs.

Table 1. The Instrument Blueprint

Variable	Indicators	Sub-Indicators	Item Numbers
Creativity in designing batik motif	a. Novelty	a. Novelty in the process, technique, material, and concept in designing batik motif.	1,2,3,4,5
		b. Creating original and germinal product.	6,7,8,9
	b. Solution	a. Product meaningfulness: the motif can be applied for practical use.	10,11,12
	c. Elaboration and synthesis	a. Complexity: Different motifs are combined into a coherent unity.	13,14,15
	Total		15

2.2 Instrument Validity and Reliability

Content validity was carried out in this study to ensure the feasibility of the instruments. The instrument validators were three lecturers of Batiking Course. The feasibility formula and the results of the assessment from the three validators are presented in Table 2 and 3.

Table 2. Instrument Feasibility Criteria

Scores	Categories
8 -15	Feasible
1- 7	Unfeasible

Table 3. The Results of the Instrument Validity Assessment

Validators	Scores	Categories
1	15	Feasible
2	15	Feasible
3	15	Feasible

The above results indicated that the instrument was feasible to be used to assess the students' creativity in designing a batik motif.

Meanwhile, the instrument reliability was assessed using Cronbach's alpha formula. Sumintoro and Widhiarso set a coefficient interval for reliability [6], as shown in Table 4.

Table 4. Reliability Levels

Coefficient Intervals	Reliability Levels
< 0.11	Useless
0.11 -0.20	Depending on the circumstances
0.21 – 0.35	Useful
> 0.35	Very Useful

The reliability calculation resulted in a coefficient of 0.525, meaning that the instrument was very useful. The target of this study was that 80% of the total number of the students were creative in designing Yogyakartan batik motif.

2.3 Data Analysis
In this classroom action research, the obtained data were analyzed using descriptive statistics. A frequency distribution was constructed by determining the interval class based on a normal curve. The level of tendency was identified using the Mean Ideal (Mi).

$$> (Mi + 1.5\ SD) = \text{Very Creative} \quad (1)$$
$$Mi \text{ up to } Mi + 1.5\ SD = \text{Creative} \quad (2)$$
$$Mi \text{ up to } (Mi - 1.5\ SD) = \text{Less Creative} \quad (3)$$
$$< (Mi - 1.5\ SD) = \text{Not Creative} \quad (4)$$

3. Result and Discussion

3.1 A Description of the Research Implementation
This study was carried out in three meetings. The students were already given a prior explanation regarding the issues related to the Batiking Course, especially about the types of batik motifs: primary, secondary, or *isen-isen*.

3.2 The Implementation of Resource-Based Learning
3.2.1 Pre-Cycle
3.2.1.1 Planning. The study began with a coordination among the researcher team (the course lecturers and the researchers), discussing the research materials, grouping, RBL implementation plan, and instrument development. This discussion resulted in the following agreement: (1) In developing the Yogyakartan batik motif, the students did a stole making. There were 8 groups consisting of 5 students each; (2) The learning model implemented by the course lecturer was the conventional one; (3) Research instruments were developed to evaluate the process and results of the research.

3.2.1.2 Implementing and Observing. In the implementation phase, the students were not grouped. They were observed when doing some exercises; (1) Introduction: The lecturer described the learning objectives, motivated the students by asking some questions about batik motifs, and gave a brief description of the subject; (2) Main Activities: Using a conventional teaching model, the lecturer delivered the learning materials and gave a job sheet to the students. Then he/she gave some examples about how to design primary, secondary, and *isen-isen* motifs. At the end, the students were instructed to develop a motif that would be implemented in a stole; (3) Closing: The students submitted their work, and the lecturer gave comments about it.

3.2.1.3 Reflecting. The observation results indicated the following problems: (1) many students could not develop a motif yet; (2) many students were not confident of their work; (3) the classroom was quiet at the beginning of the task assignment, but noisy at the end; (4) the assessment results showed that many students did not do the task seriously.
Based on the reflection results, the conventional learning model was then substituted with Resource-Based Learning.

3.2.2 Cycle I
3.2.2.1 Planning. This phase focused on the preparation of everything needed in this study, including the research issues, research design, and the technical administration. In general, the activities integrated in this preparation stage included preparing the learning resources and developing the research design.

3.2.2.2 Implementing and Observing. This phase was the implementation of observational learning method using printed materials such as books and magazines, and at the same time it was also an assessment stage of the learning activities. In details, these are the observed activities: (1) the lecturer gave a brief explanation about the new learning method and stated the learning objectives to be achieved through that method; (2) the lecturer delivered the whole materials and gave a job sheet to the students; (3) the students were divided into 5 groups; (4) each group was instructed to design a batik motif to be applied in a skirt by referring to Yogyakartan batik motifs; (5) each group did the task and created a way to make the desired design; (6) at the end, each group presented their work in front of the class.

3.2.2.3 Reflecting. From the learning process in cycle 1, the following reflection can be drawn: (1) some students could not develop a motif yet; (2) some students were not confident of their work; (3) the classroom was quieter, and the discussion could start to run; (4) the groups were not so heterogeneous that there were groups of which works were already good and there were groups of which works were still not good enough; (5) after being assessed, half of the students were able to develop a motif; (6) the learning resources were still limited to printed materials so that the students' creativity could not be developed maximally.
Based on the reflection results, the existing weaknesses were to be eliminated and the implementation of Resource-Based Learning was to be maximized.

3.2.3 Cycle II
3.2.3.1 Planning. After analyzing the reflection results from Cycle I, this phase was the very first mechanism focused on the preparation of everything needed in this classroom action research, including the things related to the subject matter, research design, and its technical administration. In this cycle, some improvements were made, like making the groups more heterogeneous and widening the learning resources to the Internet. Each group was given intensive guidance during the learning process.

3.2.3.2 Implementing and Observing. In this phase, the students' learning process were observed and their works were assessed.

3.2.3.3 Reflecting. In this cycle, some students had already been able to develop batik motifs, they were more confident of their works, the classroom atmosphere was calmer, and the discussion could run well.

Based on the results of this reflection, the research was ended for the learning process had run well, and most students had already been able to develop Yogyakartan batik motifs.

3.3 The Improvement in the Students' Creativity in Designing Yogyakartan Batik Motifs Pre-Cycle.
In this pre-cycle, the students' creativity in developing Yogyakartan batik motifs was not as expected yet. The data analysis resulted in a mean score of 29, a median of 28, and a mode of 28.

Table 5. Creativity Improvement Scores

NO	Scores	Criteria	Pre-Cycle		Cycle 1		Cycle 2	
			FREQ.	%	FREQ.	%	FREQ.	%
1	46-60	Very Creative	-	0	4	11	18	50
2	31-45	Creative	10	28	22	61	16	44
3	16-30	Less Creative	26	72	10	28	2	6
4	1-15	Not Creative	-	0	-	0	-	0
		Total	36	100	36	100	36	100

In the Pre-Cycle phase, only 10 students (28%) were creative, while the rest 26 students (72%) were less creative. Therefore, the students' creativity in developing Yogyakartan batik motifs in this phase was categorized as "less creative".

This condition urged the researchers to implement group investigation learning model. The cycles implemented in this model aimed to improve the learning quality as shown in the students' creativity in developing the motifs of Yogyakartan batik.

3.3.1 Cycle I. The results of the observation showed that there had been improvements in terms of the students' enthusiasm and interest in completing their tasks, and likewise in their mastery of the learning materials. The descriptive statistics calculation resulted in a mean score of 37, a median of 30, and a mode of 30. As shown in Table 5, the students' creativity had increased by 44% as 4 students (11%) were very creative, 22 students (61%) were creative, and 10 students (28%) were less creative.

3.3.2 Cycle II. Based on the observation undertaken in this cycle, some improvements had been achieved, like in the smoothness of the learning process, the students' enthusiasm in completing their tasks, and their interest in making a more interesting design. In addition, the students' mastery of the learning materials in this second cycle was much better than that of the first cycle. The descriptive statistics calculation resulted in a mean score of 46, a median of 48, and a mode of 48. As presented in Table 5, the students' creativity from the Pre-Cycle, Cycle I, to Cycle II had increased by 72% as 18 students (50%) were very creative, 16 students (44%) were creative, and 2 students (6%) were less creative.

3.4 Discussion.

The implementation of Resource-Based Learning in Batiking Course began with the pre-cycle of the implementation of conventional model to measure the students' creativity in developing Yogyakartan batik motifs. In the main activities, the lecturer presented the materials and gave examples, and the students were asked to develop batik motifs. After the assessment process, it was concluded that the learning implementation was less effective. It was shown by the fact that many students could not develop the batik motifs yet, some students were not confident of their work, the classroom atmosphere was calm at the beginning but noisy at the end, and most students' works were not seriously made.

By considering those reflection results, the conventional learning model was replaced by Resource-Based Learning. In Cycle I, the students were exposed to such resources as books and magazines related to batik motifs. In the materials presentation phase, the lecturer gave a brief description of RBL, including the intended learning objectives. The lecturer presented the whole materials and distributed a job sheet to the students. The students were divided into five groups. Each group was given a case to solve: developing a Yogyakartan batik motif to be applied in a skirt. Each group worked together to find a way to develop the desired design. After they finished developing their motif, they had to present their work in front of the class. Through this learning model, there had been some improvements: some students had not been able to develop a batik motif yet, many students were not confident of their works, the classroom atmosphere was calmer, the discussion started to run well, the groups were not heterogeneous, and the learning resources were limited to printed materials.

This unsatisfactory result urged the researchers to continue the study to Cycle II and add the Internet as an additional learning resource. The class began with an explanation on how to use the Internet to develop a motif. The lecturer then distributed a job sheet to the students and gave a brief explanation about the motifs suitable for a dress. In this cycle, the learning process was similar to that of the first cycle. However, for an improvement, the Internet was used for reference and the groups were divided heterogeneously. In this cycle, some students had been capable of developing batik motifs and were more confident of their works. Moreover, the learning atmosphere was calmer, and the discussion has run well. More than half of the students could develop a motif well. This cycle had resulted in the desired improvements so that the study was discontinued.

In the pre-cycle phase, only 28% of the students were creative in designing the Yogyakartan batik motif. Most students produced common and unartistic motifs with uninteresting color compositions and through improper procedures. The students had difficulties in differentiating secondary motif from *isen-isen*. Susanto defines secondary motif as a motif that completes the primary motif, while *isen-isen* is a motif that fills in the empty field in the cloth [3].

For this condition, the researchers found it necessary to implement Resource-Based Learning in order to improve the learning quality as seen the students' creativity in developing Yogyakartan batik motifs. In the first cycle, 4 students (22%) were very creative, 22 students (28%) were creative, and 10 students (72%) were less creative. Therefore, in average the students were already in "creative" category. This improvement could be seen in the unique and artistic motifs they produced. The color composition they used was better and they had developed the motifs in a proper procedure. However, the students still had a problem in giving the right title for the motif.

To achieve better improvements, the researchers continued the study to Cycle II. In this cycle, as many as 18% of the students (50%) were very creative, 16 students (44%) were creative, and 2 students (6%) were less creative. From the Pre-Cycle to Cycle II, an improvement in the students' creativity as much as 72% was achieved. The motifs they developed were unique and artistic; the color composition they applied was harmonious; the title matched the motif; and the motifs were developed through a proper procedure. This study has shown that Resource-Based Learning can improve the students' creativity in developing Yogyakartan batik motif.

4. Conclusion

The implementation of Resource-Based Learning in Batiking Course bagan with Introduction stage: the lecturer gave a brief description of RBL and stated the learning objectives to be achieved through this model. In the Main Activities stage, the lecturer presented the materials, showed how to make use of the available learning resources, divided the class into several groups, assigned a task for the students, and asked the students to develop a batik motif to be presented in front of the class. Finally, the Closing stage was for evaluating the students' works as well as the learning process.

In the Pre-Cycle, Cycle I, and Cycle II phases, the mean scores of the students' creativity in developing Yogyakartan batik motif were 29 (less creative), 37 (creative), and 47 (very creative) respectively. Therefore, the number of creative students in this implementation had increased from only 28% to 94%.

Acknowledgments

In implementing Resource-Based Learning, the students should be given freedom to make use of any learning resources. Nature can also be used as a learning resource to improve the students' creativity in designing a batik motif. To encourage the students to improve their creativity in developing batik motifs, the students should be provided with a lot of tasks and exercises in this field.

5. References

[1] Utami, Munandar, S.C. (1987). Mengenal Bakat dan Kreativitas Anak Sekolah: Penuntun Bagi Guru dan Orang Tua. Jakarta: Gramedia.

[2] Prasetyo, Anindito. (2010). Batik Karya Agung Warisan Budaya Dunia. Yogyakarta: Pura Pustaka.

[3] Susanto, S.K. Sewan. (1973). Seni Kerajinan Batik Indonesia. Jakarta: Balai Penelitian Batik dan Kerajinan. Lembaga Penelitian dan Pendidikan Industri. Departemen Perindustrian.

[4] Seel, Norbert M. (2012). Encyclopedia of the Science of Learning. London: Springer New York Doordrech Heldelberg.

[5] Nasution. (2003). Berbagai Pendekatan Dalam Proses Belajar dan Mengajar. Jakarta: PT. Bina Aksara.

[6] Sumintono, B and Widhiarso, W. (2013). Aplikasi Model Rasch untuk Penelitian Ilmu-ilmu Sosial. Cimahi: Trim Komunikata Publishing House.

Video based e-module development by using smartphone for programmable logic controllers practical learning in vocational high school

Sukir[1]

[1]Faculty of Engineering, Yogyakarta State University, Yogyakarta, Indonesia

E-mail: sukir@uny.ac.id

Abstract. The purposes of this research are: (1) Develop an interactive e-module that contain videos by using smartphone; and (2) to find out the feasibility of the results of the development of an interactive video e-module by using smartphone for PLC practical learning in Vocational High School. Type of research conducted through is research and development referring to ADDIE development model according to Robert Maribe Branch. Research steps conducted are: Analyze, Design, Develop, Implement, and Evaluate. Instruments used are questionnaire and observation sheet. Data collected is analyzed descriptively. Research result indicates: (1) An interactive video e-module by using smartphone for PLC practical learning in Vocational High School has been produced, which installed in smartphone with the name of Basic PLC Module application; and (2) Video Based Interactive E-module by using smartphone has the value of 80,75% which means it can be considered feasible to be used in PLC practical learning in Vocational High School.

1. Introduction

Today science and technology development that applied in industrial world is highly advanced. A notable example in science and technology in electrical engineering that applied in industry is Programmable Logic Controller (PLC). PLC used in industry from time to time is more advanced and sophisticated, which serve as brain to control various tool or machinery in industry. PLC application example in industry is: PLC usage as lift controller in building; conveyor belt control system; potable water bottle filling; car chassis welding and pressing; industry product sorting controller; etc.

Based on PLC usage advancement as machines control in industry as mentioned above, Vocational High School especially on Competence of Expertise in Installation of Electricity Utilization, which is an educational institution graduaate producer fo industry workforce, must teach the students PLC competence. PLC competency for student is ver important, so after graduation student is proficient in PLC, and capable to handle works relevant to PLC technology in industry [1,2,3].

Based on study conducted in Vocational High School 1 Sedayu, Bantul, Vocational High School 2 Pengasih, Kulon Progo, and Vocational High School 2 Wonosari, Gunung Kidul, in implemening 2013 curriculum for Competence of Expertise in Installation of Electricity Utilization, obtained data PLC competency that taught for grade XII in the 5th in the form of PLC practical learning of electrical motor installation subject. In PLC practical learning of electrical motor installation subject, it is found that student PLC competency in all of 3 Vocational High School is considered low. It is indicated by student report data in all of 3 Vocational High School on academic year of 2015/2016, 2016/2017, and 2017/2018 on average student that attain PLC competency score above 70 is still under 60%.

During preliminary study obtained data on all 3 Vocational High School, PLC practical learning done by using practical tool such as industry grade trainer kit which are: Omron, Zelio, Festo and Siemens, small with average size 35 cm x 25 cm x 15 cm, approximately as big as a tote bag. The usage of small PLC trainer kit for practical learning has proven that the trainer kit is not ergonomic. All PLC practical learning conducted in a group of students summed 3-4 person each group with a PLC trainer kit, practical learning done through this method deem not effective. It is shown by the facts that (1) only 1 or 2 student in front of trainer kit and capable to perform a good practical learning, the rest seems only watch their peers; (2) between students can be seen fighting over trainer kit; (3) passive student can be seen because students do not have the chance to use trainer kit; (4) practical learning atmosphere is uncomfortable (5) often student with high ego use PLC trainer kit selfishly, etc. It seems industry grade trainer kit only feasible for individual practical learning, not suitable for group practical learning. Due to limitation of PLC trainer kit and funding of Vocational High School to procure one, so PLC practical learning must be conducted in a group.

Moreover, to use PLC trainer kit during learning process is supported by printed material in the form of job sheet. According to interview done with the teachers, the job sheet is used since a long time ago without any meaningful change. This matter of course causing student less interested, not motivated, and easily bored. Even though today and upcoming future, job sheet is very open to innovates, for example in the form of electronic module or interactive e-module which loaded with video through smartphone. Today almost every single student has a smartphone in his/her possession, because the price is getting cheaper and internet network also getting wider and better. Smartphone in student possession also can be utilized to support PLC practical learning. Moreover, a part of PLC is programming, that could have installed in computer and laptop and can be used in independent learning especially simulation menu.

According to description above it is necessary to conduct a research on development of PLC trainer kit that ergonomic with the support of interactive e-module loaded with video by using smartphone in Vocational High School. The research boundary is video loaded e-module by using smartphone which is a module that created in the form of interactive e-learning especially on question based test, and loaded with trainer kit tutorial video with smartphone for android. Research product produced will be used for PLC practical learning in Vocational High School, on Competence of Expertise in Installation of Electricity Utilization.

The purpose of this research is to produce an interactive e-module loaded with video by using smartphone to support the usage of PLC trainer kit in PLC practical learning for Vocational High School. Another purpose is to find out the feasibility level of interactive e-module development loaded with video by using smartphone to support PLC trainer kit usage in PLC practical learning in Vocational High School. Interactive e-module is a teaching material in the form of electronic module or e-learning that developed interactively.

Smartphone is a type of mobile application based on software use which designed to run on mobile device. Mobile devices use in Indonesia always on the rise every year. On 2013 it is shown that 86,09% of Indonesia citizen has a phone in possession. This percentage is rising every year and there are no sign of decline. This fact is indicating that most of Indonesia citizen has already utilize mobile device. In other hand, especially in mobile application, it seems that smartphone user growth in Indonesia in 2013 has reach 70% compared to previous period. Based on that report, Android is an operating system that dominates smartphone market in Indonesia with market division as high as 59,91%.

2. Method

Research type that used in this research is research and development to produce a product in the form of an interactive e-module loaded with video by using smartphone. Development model refers to

ADDIE development model according to Robert Maribe Branch [2] Steps in this research model are: analyze, design, develop, implement and evaluate.

Technique and instrument in data collecting which used in this research: (1) on study of activity need, data collected by using questionnaire, while instrument used also used in the form of questionnaire; (2) on evaluation activity toward result of study of need, and design result, data collecting technique used is through observation, while instrument used is in the form of observation sheet; and (3) on material expert, media expert, and teacher validation toward development product, data collecting technique used is observation, while the instrument used is an observation sheet. On this research, instrument validity determined by using face validity, and content validity through expert judgement. To determine instrument reliability, especially material expert, media expert and student observation sheet Scot reliability is used. Research data collected is analyzed descriptively [3,4,5].

3. Result and Discussions

After carrying out the steps of development research that refers to ADDIE development model according to Robert Maribe Branch, then obtained development product in the form of interactive e-module loaded with video by using smartphone in PLC practical learning in Vocational High School in Competence of Electrical Installation Engineering Skills department. Development result of interactive e-module loaded with video by using smartphone on PLC practical learning in Vocational High School can be seen on smartphone with the application name of Basic PLC Module, which display can be seen on Figure 1. Testing toward work method for interactive e module loaded with video by using smartphone product in PLC practical learning in Vocational High School, done by black box testing test. Black box testing meant to test the system whether it is functional as planned. In black box testing function tests are carried out for each block of whole PLC module for smartphone application. Black box testing result is shown in full on Table 1.

To obtain feasibility level on product use of interactive e-module loaded with video by using smartphone in PLC practical learning in Vocational High School, so the product application is validated by material expert, media expert, and student. Validation activity toward product application is done by 2 material experts, 2 media experts, and 3 students which result is briefly presented in Table 2.

This developing product is in a form of interactive e-module video loaded with smartphone in PLC practical learning in Vocational High School can be seen on Smartphone with the name of Basic PLC Module. Black box testing result on interactive e-module product loaded with video by using smartphone in practical learning in vocational high school shown that 19 item of case study is perfectly successful. This means that 100% of application is functioning in PLC e-module like planned [6,7,8].

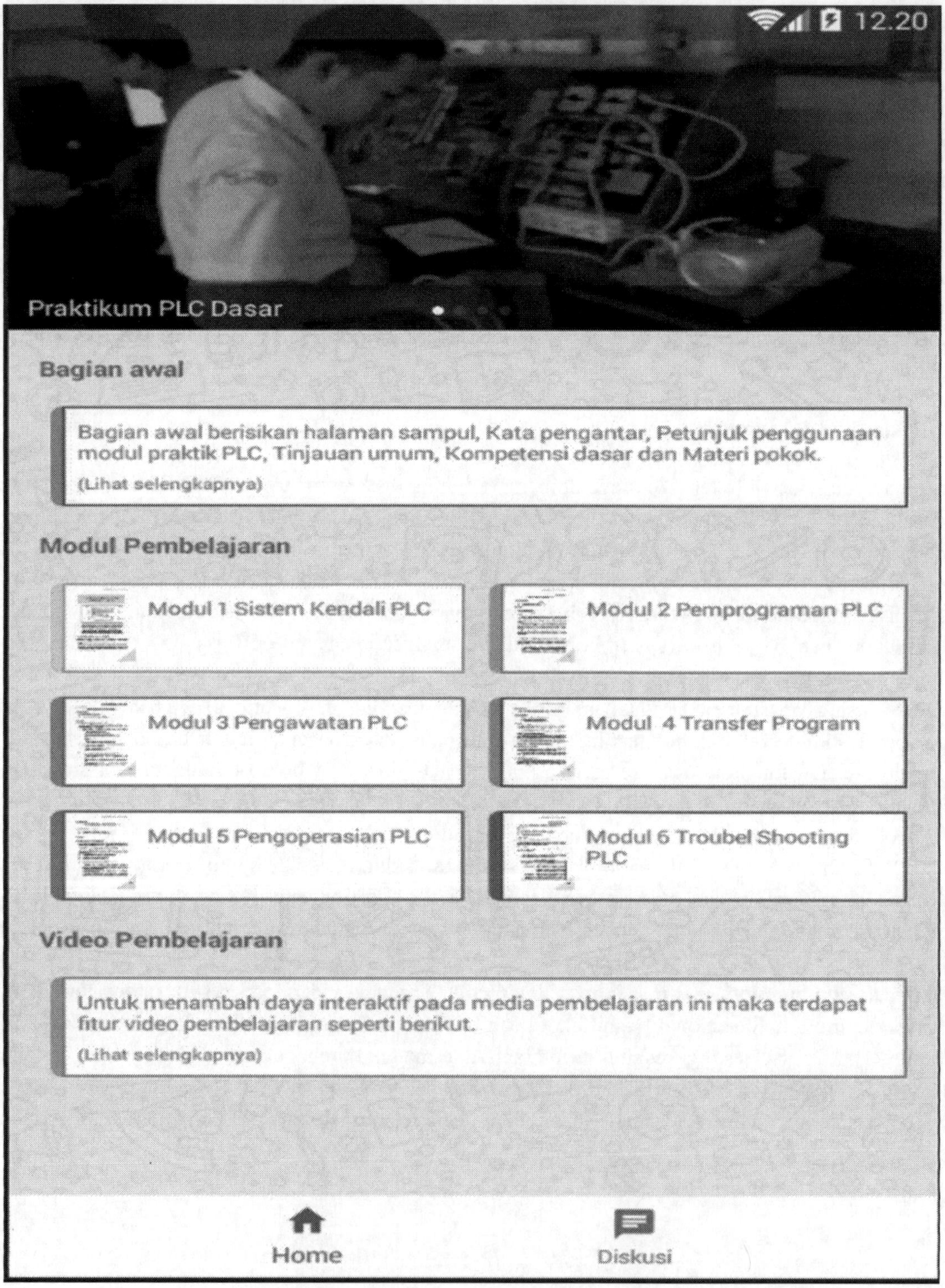

Figure 1. Display of basic e-module PLC application products on smartphones

Table 1. Black box test result on interactive e-module loaded with video by using smartphone in PLC practical learning in Vocational High School.

	Case study	Indicator	Result
1.	Install application on android smartphone.	User can install application on their own android smartphone	Succeed
2.	Opening Application	User can open application and get into main menu	Succeed
3.	Access illustrations.	Users can access illustration e-modules.	Succeed
4.	Access the beginning part.	Users can access the initial section consisting of module covers, preface, module usage instructions, module general review, basic competencies and subject matter of the PLC module.	Succeed
5.	Access to Module 1.	Users can access learning module 1 about PLC control systems	Succeed
6.	Work on formative tests 1.	Users can work on formative tests 1 and can find out the scores they achieved.	Succeed
7.	Access to Module 2.	Users can access learning module 2 about PLC programming techniques.	Succeed
8.	Work on formative tests 2.	Users can do formative tests 2 and can find out the scores they achieved.	Succeed
9.	Access to Module 3.	Users can access learning module 3 about PLC installation and wiring.	Succeed
10.	Work on formative tests 3.	Users can do formative tests 3 and can find out the scores they achieved.	Succeed
11.	Access to Module 4.	Users can access learning module 4 about inserting programs into the PLC.	Succeed
12.	Work on formative tests 4.	Users can do formative 4 tests and can find out the score they achieved.	Succeed
13.	Access to Module 4.	Users can access learning module 5 about operating a PLC control system.	Succeed
14.	Work on formative tests 5.	Users can do formative tests 5 and can find out the scores they have achieved.	Succeed
15.	Access to Module 6.	Users can access learning module 6 about tracking PLC control system errors.	Succeed
16.	Work on formative tests 6.	Users can do formative tests 6 and can find out the scores they achieved.	Succeed
17.	Access learning videos.	Users can access available PLC practice learning videos.	Succeed
18.	Asking question.	Users can ask questions through discussion forums. and stored in the database.	Succeed
19.	Exit the application.	Users can exit the application by clicking out of the application.	Succeed

Table 2. A summary of the results of the validation of material experts, media experts, and teachers on interactive e-module products using smartphones at the Practice of PLCs in Vocational Schools.

	Aspect	Material Expert		Media Expert		Student	
		Score (%)	Category	Score (%)	Category	Score (%)	Category
1.	Relevance	87,5	Very Feasible	85	Very Feasible	83,75	Very Feasible
2.	Benefit	82,5	Very Feasible	83,75	Very Feasible	83,13	Very Feasible
3.	Material	83,33	Very Feasible	82,5	Very Feasible	82,29	Very Feasible
4.	Technique	70,83	Feasible	71,87	Feasible	72,5	Feasible
Overall Aspect		81,04	Feasible	80,78	Feasible	80,42	Feasible

Based on validation result of material expert toward interactive e-module loaded with video by using smartphone in PLC practical learning in Vocational High School validity, yield result of 81,04% which is in feasible category. Validation done by media expert toward feasibility of the product scored 80.78% which is in feasible category also. The same goes for validation done by student toward product validity, obtained score of 80,42% which is also in feasible category. Based on validation result of material expert, media expert and student toward interactive e-module loaded with video by using smartphone in PLC practical learning in Vocational High School above, the mean of validity scored 80,75% which is in feasible category. With that it can be said that interactive e-module product loaded with video by using smartphone in PLC practical learning in Vocational High School fulfill feasible category to be applied in PLC practical learning in Vocational High School Competence of Expertise in Installation of Electricity Utilization.

4. Conclusion

Interactive e-module loaded with video by using smartphone for PLC practical learning in Vocational High School is produced, which is installed in smartphone with the name of Basic PLC Module. Interactive e-module loaded with video by using smartphone scored 80,75% which is in feasible category to be used in PLC practical learning in Vocational High School.

5. References

[1] Aras, M.S.M. & Azis, F.A. (2014). ROV trainer kit for education purposes. *International journal of science and research (IJSR)*. Volume 3, Issue 5, May 2014

[2] Branch, R.M. (2009). *Instruction design: the ADDIE approach.* New York: Springer Science + Business Media

[3] Guo, N. & Chen, Z. (2012). Explore of teaching practice on programmable logic controller. *International Conference on Education Technology and Computer (ICETC 2012)*, Vol.43, IACSIT Press, Singapore

[4] Hasan, S. (2006). *Analisis perakitan trainer unit berdasarkan aplikasi konsep refrigerasi pada mata kuliah Sistem Pendingin.* Diambil pada tanggal 28 September 2014, cited at http://file.upi.edu/.../JUR.../artikel_trainer_refr_1.pdf

[5] Hong, at.al. (2013). Development of an Educational Robotic Training Kit. *Advanced Science Letters.* Volume 19, Number 11, November 2013, pp. 3304-3306(3)

[6] Ranjit, at.al. (2012). Development of solar educational trainer kit. *International journal of engineering and innovative technology (IJEIT).* Volume 2, Issue 3, September 2012.

[7] Rauner, F., et. al. (2008). *Handbook of technical and vocational education and training research.* Bremen: Springer Science Business Media

[8] Regional Education Center of Science and Mathematics. (2011). *Media pembelajaran Ilmu Pengetahuan Alam.* Accessed at 30 Juli 2012, from(complong.files.wordpress.com/2011/10/bab- iii.pdf)

IOP Conf. Series: Materials Science and Engineering **535** (2019) 012019 doi:10.1088/1757-899X/535/1/012019

Automatic water filling and capping machine trainer kit programmable logic controller controller prototype for vocational high school learning validity

Sukir[1], N Yuniarti[2], S Yatmono[3], and T Wicaksono[4]

[1,2,3,4]Faculty of Engineering, Universitas Negeri Yogyakarta, Yogyakarta, Indonesia

E-mail: sukir@uny.ac.id

Abstract. The purpose of this article is to find out validity level PLC controlled automatic water filling and capping machine as trainer kit for learning in industry automation competence expertise in vocational high school. Research type used in this research is research and development (R&D) according to Robert Maribe Branch. carried out steps are; Analyze, Design, Develop, Implement, and Evaluate. Instrument used is questionnaire and observation sheet. Collected data is analyzed descriptively. Research result shown that validation toward *automatic water filling and capping machine* prototype as trainer kit for PC practical learning in vocational high school which done by media expert, material expert, and student shown result in very satisfactory category, indicated by: (1) media experts give average score 87.5% (very satisfactory); (2) material experts give average score of 86,9% (very satisfactory); and (3) student give average score of 95,3% (Very Good)

1. Introduction

Science and technology today is rapidly developing. This science and technology advancement greatly affects the industry sector. Today industry is constantly developing and demanded to optimize operation skill quickly, cost-effective and innovative. Various industries today use high and sophisticated technology to create a product with superior quality to compete globally.

According to Ministry of Industry of Republic of Indonesia, on 2018 industry is the most significant sector that will greatly enhance Indonesia's and the world economic growth, this industry is divided into 5 sectors. These 5 sectors are basic iron, food and beverages, lifting tool machine, chemistry, and electronics.

Especially in industry that move in food and beverage sector, Central Bureau of Statistic (2017) records that food and beverage industry is capable to donate gross domestic product, non-gas fuel on quarter III/2017 summing 34,95% or highest in comparison with other sectors. Another sectors like metal, computer, electronic, optic and electrical tool industry contributes 10,46% and transportation equipment industry contributes 10,11%. In other hand, Ministry of industry also records that workforce contribution is dominated by food industry totaling 3.316.186 person or 21,34%.

To improve national food and beverage industry growth and structure, Ministry of Industry of Republic of Indonesia confirms that his party is supportive so this priority sector joins and support government program in education and industry vocational training. As an effort to fulfill the needs of workforce, Mministry of Industry asked food and beverages industry to develop collaboration with vocational high school in area near their factory to improve vocational high school competence.

Industry development must be supported by human work force that is reliable and competent. Vocational high school must produce graduate that competent and capable to compete according to industrial world need and demand that constantly developing. But today based on data obtained by central bureau of Indonesian labor, the sum of workforce in August 2017 is 128,06 billion is rising into 2,62 billion in comparison in August 2016. Component forming of the workforce is citizen that works and unemployed citizen. On August 2017, 121,02 billion working citizen and 7,04 billion unemployed. Seen from education level, open unemployment rate in vocational high school is the highest in comparison with similar level of education which is 11,41 percent. In other word, there is a supply of labor that is not absorbed especially in the vocational high school level. Cited from www.suaramerdeka.com (2017), Dirjen Dikdasmen states the high employment rate of vocational high school is due to inconsistency between available needs, number of graduation, and quality or competency of vocational high school graduate that is not absorbed in the working world.

In modern industry today more usage of industrial automation system is implemented to improve productivity. Through industry automation system, complex work can be handled and controlled by using machine that work automatically with accurate result, those machine utilize *Programmable Logic Controllers* (PLC) as controlling device. Through industry automation in vocational high school in competence of industry automation, student can learn science and technology on industry automation on the study program. But till this day there are obstacle and problem in the learning process especially in PLC learning so that graduate produced is still not met the competence that is needed by industry.

Researcher conducted a preliminary study to find out situation and condition on learning process conducted in industry automation expertise program in vocational high school especially on PLC practical learning. Preliminary study is done through observation in a vocational high school that has expertise program in industry automation. Observation is done in vocational high school 2 Depok, Sleman Yogyakarta Special Regency by interview with head of industry automation expertise program. According to the case study, it is found that learning process in PLC practical learning is done through programmable control system subject. In that subject learning media is very limited and still uses conventional learning media that not suited with today industry. Funding limitation is the source of problem why learning media is so limited. Drs. Bambang Irianto gives suggestion to researcher on learning media that will be developed which is, developed product is expected to be a learning media that is applicative and suited with today industry so student will be ready in industry in the future. Moreover learning media product that is applicative need to be supported with PLC practical learning module with mental revolution material, with the hope that student will have better mental.

According to the background the research is conducted and produces a PLC controlled prototype automatic water filling and capping machine trainer kit for learning in Industrial Automation Engineering Skills Competencies in Vocational High School. The purpose of this article is to find out validity level of PLC controlled automatic water filling and capping machine trainer kit for learning in Industrial Automation Engineering Skills Competencies in Vocational High School.

Hasan stated that a form of learning media is a trainer kit, that is a set of tools in lab that sed as learning media, which is a combination between work model or work process and object model or artificial object [1]. Artificial object is a simplified structure of main part in a process or system that complex. A set of tools in lab or workshop in the form of real object model or artificial object in industry as learning media equipped with tool guide is called industrial trainer kit. Industrial trainer kit is intended to support student learning process in applying knowledge or concept received on object or tools in industry. That opinion gives meaning that industrial trainer kit is a practice tool that used in learning in lab or workshop in the form of real object or artificial object in industry equipped with job sheet or practical learning module.

A form of another media is practical tool. This matter is strengthened by Rauner that practical is a tool or lab equipment [2]. Because of that it can be said that practical tool is the same as media in the concept of meaning, which is an information vehicle of learning information that channel learning information between information source with student as study information recipient, which covers person, material, tool and activity that allow student to gain information, skill, and attitude. Another definition on practical tool is given by *Regional Education Center of Science and Mathematics*, states that practical tool is a tool or set of tools that used directly in practical learning or experiment in lab or workshop to develop skill, concept or knowledge for student [3]. According to definition above gave concept that the existence of practical tool in practical learning activities especially in vocational high school become very important, because student will use it in the practica which will give skil and needed concept mastery in working world.

Research that is relevant with this research is research done by Ranjit, on *solar educational trainer kit* development which yield result that *solar educational trainer kit* is capable to give the user an easy understanding on electricity energy charging, and conversion of DC electricity into AC electricity [4]. Another research is done by Aras and Azis on Remotely Operated Vehicle (ROV) trainer kit for education purposes, which shown result that ROV trainer kit is capable of giving an easy understanding on underwater vehicle control system that can be operated from afar [5]. Research conducted by Hong that investigating on development of educational robotic training kit, shown that the usage of educational robotic training kit in learning can greatly reduce cost, student easy to understand, and simple [6]. Other research done by Guo and Chen is on "Explore of Teaching Practice on Programmable Logic Controller", it is shows that demonstration and experiment method by using PLC Siemens S7-200 in effective learning can improve student study interest, active class atmosphere, enhance teaching interaction, and increasing whole effect of the learning process in class [7]. All of the 4 mentioned researches above are different with this research, which is the function of the working object that developed on all 4 researches is different with this research development.

2. Method

Research type used in this research is research and development (R&D) which refers to ADDIE [8]. Steps taken in this research are: Analyze, Design, Develop, Implement, and Evaluate. Technique and instrument in in collecting data needed in this research are; (1) in study of need, data is collected through questionnaire instrument; (2) in evaluation activity on analysis of need, and design result, collecting data technique used is observation, while instrument used is observation sheet; and (3) in material expert, media expert, and student validation on developed product collecting data technique used is observation, while instrument used is observation sheet. In this research, instrument validity is determined by face validity, content validity, through expert judgement, to determine the reliability of instrument especially material expert, media expert, and teacher observation sheet, used Scot reliability. Data collected through this research is analyzed descriptively.

3. Result and Discussions

After research steps and development by using ADDIE according to Robert Maribe Branch, yield result the automatic water filling and capping machine prototype trainer kit of PLC practical learning in vocational high school. As for automatic water filling and capping machine prototype trainer kit prototype in PLC practical learning in vocational high school is shown in Figure 1.

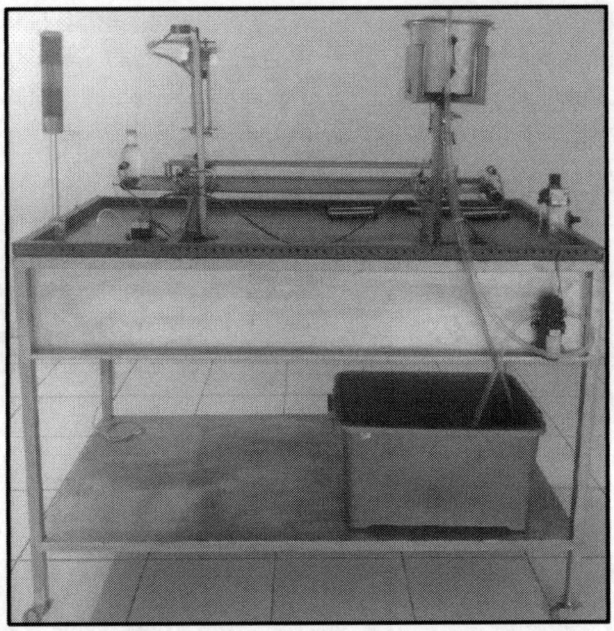

Figure 1. Automatic water filling and capping machine trainer kit prototype as PLC practical learning trainer kit for vocational high school.

To figure out the validity level of the product, the automatic water filling and capping machine prototype trainer kit for PLC learning in vocational high school, the product is validated by 2 material expert, 2 media expert, and 4 vocational high school student. Validation is done by 2 expert in media field result is briefly presented in Figure 2.

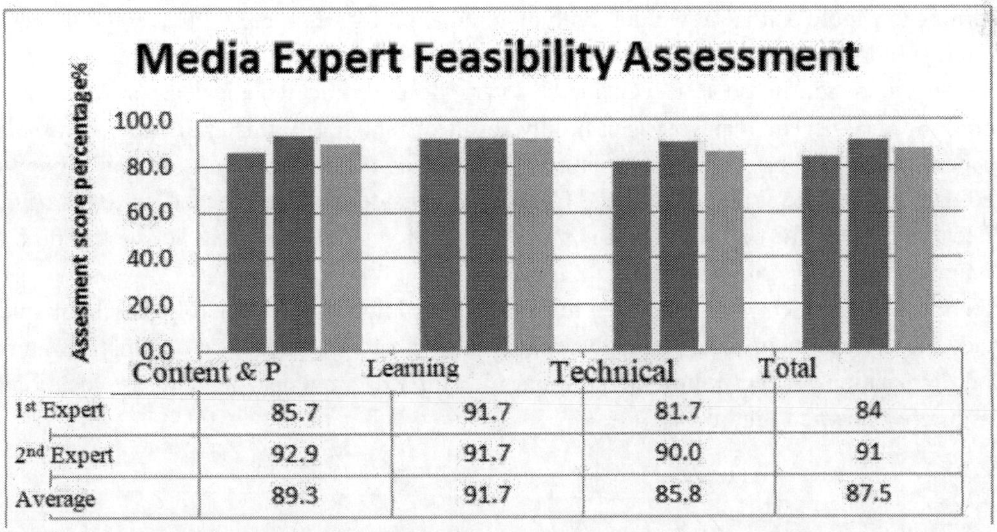

	Content & P	Learning	Technical	Total
1st Expert	85.7	91.7	81.7	84
2nd Expert	92.9	91.7	90.0	91
Average	89.3	91.7	85.8	87.5

Figure 2. Feasibility Assessment Chart of Product Media Experts

Figure 2 shows that both experts give score with average percentage in content and purpose as much as 89.3% so it is considered in "very decent" category, learning aspect as much as 91,7%, so it is considered in "very decent" category an in technical aspect 85.8% so it is considered in "very decent" category also. Based on data collected Automatic Water Filling and Capping Machine is considered "very decent" in terms of media as a PLC learning media to be used in Programmed system control in vocational high school with Industrial Automation Engineering Skill expertise competency. Validation done by 2 material experts on the product is briefly presented in Figure 3.

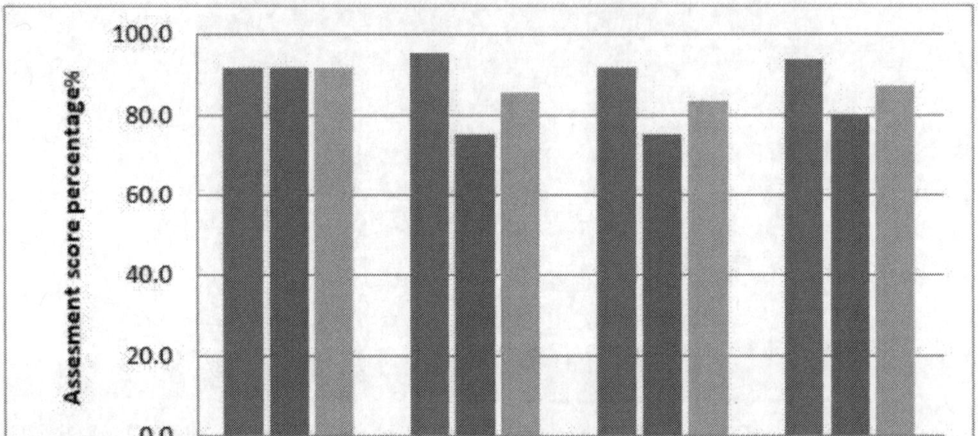

	Content & P	Learning	Technical	Total
1st Expert	91.7	95.5	91.7	93.75
2nd Expert	91.7	75.0	75.0	80
Average	91.7	85.2	83.3	86.875

Figure 3. Material Expert Feasibility Assessment Chart

According to Figure 3 it is shown that both material expert gives score with average percentage in content and purpose as much as 91,7% so the product is considered in "very decent" category, learning aspect as much as 85.2%, so it is considered "very decent" and technical aspect 83.3% so it is considered in "very decent" category. Feasibility result of material by material expert is obtained by average percentage of every aspect summing 86,9% so it is considered "very decent", based on the collected data, Automatic Water Filling and Capping is considered "very decent" in terms of material as PLC learning media to be used in Programmed control system subject in vocational high school with Competency in Industrial Automation Engineering.

Furthermore, the student also requested to give their opinion as vocational high school student as user candidate. Assessment result through vocational high school student on automatic water filling and capping machine product prototype as trainer kit for PLC practical learning in vocational high school is briefly shown in Figure 4.

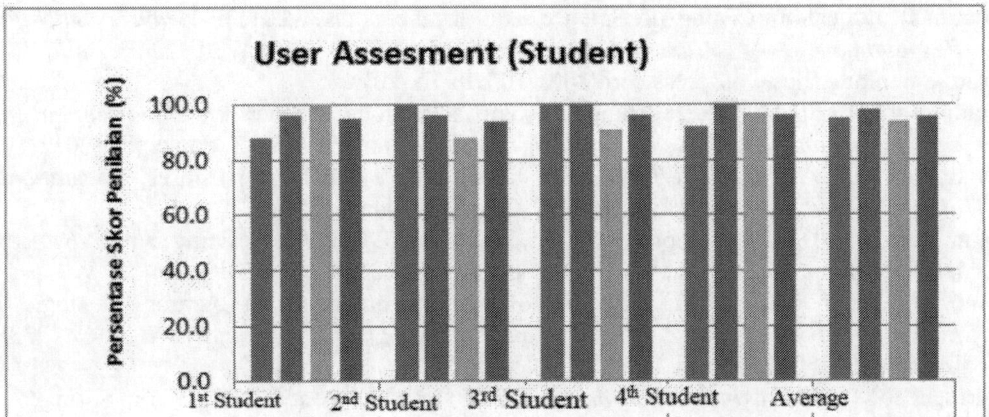

	1st Student	2nd Student	3rd Student	4th Student	Average
Content & P	87.5	100	100	91.7	94.8
Learning	95.8	95.8	100	100	97.9
Technical	100	87.5	90.6	96.9	93.8
Total	95	93.8	96.3	96.3	95.3

Figure 4. Students Assessment graphic.

Based on Figure 4 it is shown that al 4 student give score with average percentage in content and purpose aspect as much as 94,8% so it is considered in very decent" category. Learning aspect as much as 97,9% so it is considered in "very decent" category and technical aspect as much as 93.8% so it is considered in "very decent" category. Through user test result it is found that average percentage of every aspect in the product assessment score 95,3%. So it is considered in "very decent" category. Based on collected data, Automatic Water Filling and Capping Machine prototype is considered "very decent" as PLC learning media in programmed control system subject in vocational high school in Competency in Industrial Automation Engineering expertise.

4. Conclusion

Validation on Automatic Water Filling and Capping Machine prototype as trainer kit for PLC practical learning in vocational high school is done by media expert, material expert, and student shown satisfactory result, this is indicated by; (1) validation is done by media expert on this product scored 85,75% on average which means it is considered in "very decent" category; (2) validation done by material expert on this product scored 86,9% on average which means this product is considered "very decent" and' (3) Assessment done by student on product scored 95,3 in overall aspect so it is considered in "very decent" category.

5. References

[1] Hasan, S. (2006). *Analisis perakitan trainer unit berdasarkan aplikasi konsep refrigerasi pada mata kuliah Sistem Pendingin.* Diambil pada tanggal 28 September 2014, cited at http://file.upi.edu/.../JUR.../artikel_trainer_refr_1.pdf

[2] Rauner, F., et. al. (2008). *Handbook of technical and vocational education and training research.* Bremen: Springer Science Business Media

[3] Regional Education Center of Science and Mathematics. (2011). *Media pembelajaran Ilmu Pengetahuan Alam.* Accessed at 30 Juli 2012, from(complong.files.wordpress.com/2011/10/bab- iii.pdf)
[4] Ranjit, at.al. (2012). Development of solar educational trainer kit. *International journal of engineering and innovative technology (IJEIT).* Volume 2, Issue 3, September 2012
[5] Aras, M.S.M. & Azis, F.A. (2014). ROV trainer kit for education purposes. *International journal of science and research (IJSR).* Volume 3, Issue 5, May 2014
[6] Hong, at.al. (2013). Development of an Educational Robotic Training Kit. *Advanced Science Letters.* Volume 19, Number 11, November 2013, pp. 3304-3306(3).
[7] Guo, N. & Chen, Z. (2012). Explore of teaching practice on programmable logic controller. *International Conference on Education Technology and Computer (ICETC 2012),* Vol.43, IACSIT Press, Singapore
[8] Branch, R.M. (2009). *Instruction design: the ADDIE approach.* New York: Springer Science + Business Media

Model development for garment design assessing using DEMATEL

M A Jerusalem[1]

[1]Engineering Faculty, Yogyakarta State University, Yogyakarta, Indonesia

E-mail: adam_jerusalem@uny.ac.id

Abstract. Various models for evaluating the best garment design have been developed, however most of them disregard interdependence among their elements. This study aims to develop comprehensive model that consist of intrinsic and extrinsic elements for assessing garment design schemes, and take into consideration of interrelation between its elements. The Decision-Making Trial and Evaluation Laboratory (DEMATEL) was utilized for analysing a relationship between the key components for assessing garment design. The research constructed the garment design evaluation framework that consider intrinsic and extrinsic elements and interdependence elements. The model can be used as a framework for evaluating garment design with some various advantages: considering interrelations between components and influences between elements.

1. Introduction

To meet consumer needs for design and aesthetics, garment designers must consider various factors in designing clothing such as garment trends, culture, social norms, time and place [1]. Designing a garment design in a garment industry in which the production process is mass production like Zara is one of the complex decision-making processes for a company. Choosing a combination of materials, colours, patterns for the right target consumers, and other attributes need to involve judgments from experts and garment designers. Therefore, developing a multi-attribute decision making approach to evaluate garment design plays an important role in the field of clothing, especially the garment industry.

Some studies discussed how to evaluate garment design and it were mostly focus on elements and methods for assessing garment design. For example, Lin and Twu proposed eight evaluation elements within the framework of their decisions: themes, best seller modifications, new ideas, garment forecasting, garment events, opponent abilities, brand image, and product position [2]. Three elements - fabric, colour, and style - for analyzing garment design have been implemented in research of Goncu and Bayazit [1]. Harr suggested a series of different elements including time management, concepts, figure development, development of garment designs and interactions in images, rendering, image relationships, composition, quality, style, and presentation [3]. Kidd and Workman developed an evaluation model consisting of design creativity, aesthetic appeal, functionality, suitability and originality [4]. The elements discussed are intrinsic factors in garment design. On the other hand, Jones concentrated on extrinsic factors in garment design such as marketing when evaluating garment designs. The elements include the age of the consumer, gender, demography, lifestyle, physical characteristics, psychographics, social class, values and attitudes, economic conditions, and religion [5].

There are a number of studies that not only considered the elements in evaluating garment designs, but also developed approaches to choosing the ideal or best garment design alternative. Goncu and Bayazit used the AHP method [1]; Lin and Twu applied fuzzy AHP [2]; Lin compared AHP method with consistent fuzzy preference relation [6]; Lin used method of Fuzzy Delphi and Fuzzy AHP [7]; and Lin and Twu used Fuzzy AHP and Technique for Order Preference by Similarity to Ideal Solution (TOPSIS) [2] . Generaly, these studies do not consider the existence of interdependencies or relationships between existing elements. Even though in the real world, mutual relations between elements are a necessity.

Table 1. State of the art of research on garment design

Research	Method	Objective
Goncu and Bayazit, 2007, [1]	AHP	Looking for garment trends
Dadashian et al., 2009, [8]	New hierarchical method	Quality assessment of textile products
Lin et al. [9]	Gray model	Prediction of garment colour trends
Lin and Twu, 2012, [2]	Combination of a fuzzy AHP and TOPSIS	Evaluate garment designs
Lin and Twu, 2012, [10]	Fuzzy AHP	Choose the best choice among garment trends
Lin, 2013, [7]	Fuzzy Delphi method and fuzzy AHP	Evaluate garment designs
Lin, 2014, [6]	Consistent fuzzy preference relation	Choose the best garment design

One important issue in the previous studies is that elements are assumed to be independent of each other. This means that there is no dependency or relationship between one element and other element. This independent situation does not reflect the real world situation. This condition motivates the author to conduct this study by developing a model for evaluating garment designs that considers the existence of interrelationships between groups and elements. Therefore, This study aims are: 1) to compile comprehensive elements for evaluating garment designs; 2) to develop a model for decision analysis that considers the interdependence between evaluation groups and the influence between the elements for assessing several garment scheme alternatives. In order to analize its relationship, this study implement DEMATEL.

DEMATEL can be used to identify a relationship of network structure, the level of interdependence, and a vital dimensions of a scheme [11]. DEMATEL was developed by The Battelle Memorial Institute through its Geneva Research Centre. The DEMATEL is useful tool for composing the interrelation between factor or elements in order to create the impact of a network relationship maps (NRM). Its NRM can visualize the structure of complicated causal relationships with matrices or diagraphs. By using the DEMATEL, the relation between elements and causal dimension of a complex system can be converted to a comprehensible structure model [11]. DEMATEL can identify the network relationship structure and vital influential component of a garment scheme can be obtained through implementing this method. The steps of DEMATEL are as follows [11].

Step 1. *Decide the initial average matrix by score.* The level of direct influence between components is calculated using a scale of 0 to 4. These numerical values can be calculated to be an average matrix, ***A***, as follows:

$$A = \begin{bmatrix} a_{11} & \cdots & a_{1j} & \cdots & a_{1n} \\ \vdots & & \vdots & & \vdots \\ a_{i1} & \cdots & a_{ij} & \cdots & a_{in} \\ \vdots & & \vdots & & \vdots \\ a_{n1} & \cdots & a_{nj} & \cdots & a_{nn} \end{bmatrix} \tag{1}$$

Step 2. *Decide the initial influence matrix.* The normalization of the average matrix, ***A***, will produce the initial direct influence matrix, $X(X=[x_{ij}]_{nxn})$. Based on calculations of Formulas (2) and (3), can be obtained matrix ***A***.

$$X = s.A \quad (2)$$

$$s = \min\left[1/max_i \sum_{j=1}^{n} |a_{ij}|, 1/\ max_j \sum_{i=1}^{n} |a_{ij}|\right] \quad (3)$$

Step 3. *Conclude the full influence matrix.* The indirect influence between different component can be decided by considering the powers of matrix ***A***. An aggregation of the direct effect and all of the indirect effects will produce the total influence matrix, ***T***, as formula (4).

$$T = X + X^2 + \ldots + X^k = X(I-X)^{-1} \quad (4)$$

where $T=[x_{ij}]_{nxn}$ and $(I-X)(I-X)^{-1}=I$. Each row and column sum of total matrix ***T*** can be obtain after getting the total influence matrix.

$$r = (r_i)_{nx1} = \left[\sum_{j=1}^{n} t_{ij}\right]_{nx1},\ c = (c_j)_{nx1} = \left[\sum_{i=1}^{n} t_{ij}\right]'_{1xn} \quad (5)$$

where the superscript ' indicates transpose.

Step 4. *Decide the network relation map.* This study implements a threshold value, α, to eliminate unimportant effect in full influence matrix ***T***. If the elements' values in matrix ***T*** are less than α, then the values are zero. The α-cut step generates a new matrix, the α-cut total influence matrix, T_α. Therefore, the network will simpler.

$$T_\propto = \begin{bmatrix} t_{11}^{\propto} & \cdots & t_{1j}^{\propto} & \cdots & t_{1n}^{\propto} \\ \vdots & & \vdots & & \vdots \\ t_{i1}^{\propto} & \cdots & t_{ij}^{\propto} & \cdots & t_{in}^{\propto} \\ \vdots & & \vdots & & \vdots \\ t_{n1}^{\propto} & \cdots & t_{nj}^{\propto} & \cdots & t_{nn}^{\propto} \end{bmatrix} \quad (6)$$

After normalizing the α-cut total influence matrix, we obtain the network relation map based on the normalized matrix, T_s, as follows.

$$T_s = \begin{bmatrix} t_{11}^{\propto}/d_1 & \cdots & t_{1j}^{\propto}/d_1 & \cdots & t_{1n}^{\propto}/d_1 \\ \vdots & & \vdots & & \vdots \\ t_{i1}^{\propto}/d_2 & \cdots & t_{ij}^{\propto}/d_2 & \cdots & t_{in}^{\propto}/d_2 \\ \vdots & & \vdots & & \vdots \\ t_{n1}^{\propto}/d_3 & \cdots & t_{nj}^{\propto}/d_3 & \cdots & t_{nn}^{\propto}/d_3 \end{bmatrix} = \begin{bmatrix} t_{11}^{s} & \cdots & t_{1j}^{s} & \cdots & t_{1n}^{s} \\ \vdots & & \vdots & & \vdots \\ t_{i1}^{s} & \cdots & t_{ij}^{s} & \cdots & t_{in}^{s} \\ \vdots & & \vdots & & \vdots \\ t_{n1}^{s} & \cdots & t_{nj}^{s} & \cdots & t_{nn}^{s} \end{bmatrix} \quad (7)$$

where $d_i = \sum_{j=1}^{n} t_{ij}^{\propto}$. (8)

2. Method

This study uses a quantitative research approach where computation follows the computational algorithm method used, DEMATEL. DEMATEL is used to analyze interrelations between components of clothing design elements. The characteristics of DEMATEL which do not require a comparison of all elements against each of the other elements makes the analysis of the relationship between the elements simpler.

Ten experts were selected to give their perceptions of the interactions between the elements and the assessment of several design alternatives. The selection of experts is based on non-probability sampling techniques, namely purposive sampling [12]. These experts were drawn from academics and practitioners that have experience and competence on garment design, pattern making, garment assessment, textile, garment mass production, garment trend, garment forecasting, marketing, and customized garment product designs.

3. Result and Discussion

3.1. Group and Element

This study successfully developed the garment design evaluation model that considers intrinsic and extrinsic factors in the garment design process. The model consists of groups and elements. The essential group in evaluating garment designs consist of creation (innovation), mode (style), tone (colour), marketing, and textile (fabric). Each group has two elements (see Table 2).

Table 2. The objective, group, and element of the garment design evaluation model.

Objective	Group	Element
Evaluating garment design alternatives.	Creation	1. garment forecast 2. freshly concept
	Mode	1. suitability of the style 2. mix and match with other garments
	Tone	1. daily use hormony 2. body type and skin colour
	Marketing	1. customer understanding 2. customer lifestyle
	Textile	1. ease of care 2. textile motif

3.2. DEMATEL based Network Relations Map

Average matrix (*A*) derived from expert judgment must be obtained first as in Table 3. The result of calculation matrix *A* is then used to find out the matrix of direct influence (***D***) value.

Table 3. Matrix ***A***

Group	Creation	Mode	Tone	Marketing	Textile
Creation	0	3	2.5	2.4	2.2
Mode	2.8	0	2.7	2.3	2.5
Tone	2.6	2.3	0	1.9	1.9
Marketing	2.6	2.7	2.5	0	1.9
Textile	2.1	2.4	2.4	2	0

By using formula of $\boldsymbol{X} = \boldsymbol{s.A}$, the matrix ***D*** can be obtained (see Table 4).

Table 4. The matrix ***D***

Group	Creation	Mode	Tone	Marketing	Textile
Creation	0	0.3	0.2	0.2	0.2
Mode	0.3	0	0.3	0.2	0.2

Tone	0.3	0.2	0	0.2	0.2
Marketing	0.3	0.3	0.2	0	0.2
Textile	0.2	0.2	0.2	0.2	0

The formula (4) was implemented to get the matrix of total direct-influence, ***T*** (see Table 5).

Table 5. The matrix ***T***

Group	Creation	Mode	Tone	Marketing	Textile
Creation	2.3	2.6	2.5	2.3	2.2
Mode	2.6	2.4	2.6	2.3	2.2
Tone	2.3	2.3	2.6	1.9	1.9
Marketing	2.5	2.5	2.5	1.9	2
Textile	2.3	2.3	2.3	1.9	1.8

Based on Table 5, the DEMATEL succesfully deliver an important information that among groups and elements for assessing garment design have interrelation. Matrix ***T*** inform level of influence from each group. In other words, we can know which group has highest influnce to other. Hence, a decision maker should take it into consideration when evaluating designs of garment. In order to minimize the effects that are not too significant, the threshold value in the matrix ***T*** is set at 2. Therefore, only among marketing-tone-and textile have one-direction of relation from marketing to other element. It means that the interrelations between these element were inconsequential. On the other hand, creation, mode, and tone are influence each other. This study showed creation is the most dominan group, followed by mode and tone. This finding suggest to designer and decision maker in garment design compony must pay more attention on creation's elements first, then others. It is because creation give more affect to other.

4. Conclusion

This research has succeeded in producing a model for analyzing alternatives of garment design. This model consists of intrinsic and extrinsic component in carrying out garment designs. Intrinsic component consists of creation, tone, mode and fabric. While extrinsic component includes matters that related to marketing. In addition, this study produced a framework that considered the relationships between groups and elements where this was not considered by other studies. Through this model, designer or decision maker in the garment company can use the results and suggestions in this study to support their daily work.

5. References

[1] G. Goncu and N. Bayazit, "An application of AHP in the selection of fashion trends for specific target groups," *Des. Princ. Pract.*, vol. 1, no. 3, pp. 573–584, 2007.
[2] C. Lin and C. Twu, "Combination of a fuzzy analytic hierarchy process (FAHP) with the Technique for Order Preference by Similarity to Ideal Solution (TOPSIS) for fashion design scheme evaluation," *Text. Res. J.*, vol. 82, no. 10, pp. 1065–1074, 2012.
[3] S. J. Haar, "Development of assessment rubrics for fashion design projects," in *61st Annual Meeting of the International Textile and Apparel Association*, 2004.
[4] L. K. Kidd and J. E. Workman, "Assessment of Creativity in Apparel Design," *Cloth. Text. Res.*

J., vol. 17, no. 1, pp. 58–64, 1999.

[5] S. J. Jones, *Fashion Design*. New York: Watson-Guptill, 2002.

[6] C. Lin, "Application of CFPR to fashion design scheme selection," *Int. J. Cloth. Sci. Technol.*, vol. 26, no. 4, pp. 316–329, 2014.

[7] C. Lin, "Application of fuzzy Delphi method (FDM) and fuzzy analytic hierarchy process (FAHP) to criteria weights for fashion design scheme evaluation," *Int. J. Cloth. Sci. Technol.*, vol. 25, no. 3, pp. 171–183, 2013.

[8] F. Dadashian, M. A. S. Monfared, and H. G. Nasrabadi, "Design of a new quality assessment system for textile products," *J. Text. Inst.*, vol. 100, no. 7, pp. 640–648, Sep. 2009.

[9] J. J. Lin, P. T. Sun, J. J.-R. Chen, L. J. Wang, H. C. Kuo, and W. G. Kuo, "Applying gray model to predicting trend of textile fashion colours," *J. Text. Inst.*, vol. 101, no. 4, pp. 360–368, Mar. 2010.

[10] C. Lin and C. H. Twu, "Fuzzy MCDM for evaluating fashion trend alternatives," *Int. J. Cloth. Sci. Technol.*, vol. 24, no. 2/3, pp. 141–153, Jun. 2012.

[11] G.-H. Tzeng and J.-J. Huang, *Multiple attribute decision making: methods and applications*. Boca Raton, FL: CRC press, 2011.

[12] M. Saunders, P. Lewis, and A. Thornhill, *Research Methods for Business Students*, Seventh Ed. Essex: Pearson Education Limited, 2016.

IOP Conf. Series: Materials Science and Engineering **535** (2019) 012021 doi:10.1088/1757-899X/535/1/012021

Water level control based fuzzy logic controller: simulation and experimental works

M Khairudin[1], AD Hastutiningsih[2], THT Maryadi[3], and HS Pramono [4]

[1,3,4]Electrical Engineering Education Dept., Universitas Negeri Yogyakarta, Yogyakarta, Indonesia
[2]Civil Engineering Education Dept., Universitas Negeri Yogyakarta, Yogyakarta, Indonesia

E-mail: moh_khairudin@uny.ac.id

Abstract. This paper presents an automatic control system for Water Level Control (WLC) using Fuzzy Logic Controller (FLC). The use of FLC on the WLC system so that it can work automatically regulates the rotation speed of the motor (water pump). This system uses the input of Light-Dependent Sensor (LDR) as a light detector sensor outside space and ultrasonic detection sensors as the height of the water. As for the output of the motor rotation be as a replacement water pump for filling water in the reservoir of the water automatically. The working principle of this system is in a light sensor to detect outdoor LDR and Ultrasonic Sensor obstructed water level then the sensor will provide the input signal to the microcontroller in the form of Analog data when it is converted into a digital form signal using an Analog to Digital Converter (ADC). The data is processed using the FLC algorithm. The results show that the lighter in the outdoors and higher low water, the water-pumping Motor speed the faster or maximal. As for when the high water is medium and a bright light or daylight then the speed of motor rotation is medium and at the time of the highest water is maximal then the motor will die. Conversely, if the light on the outdoor is darker or at night and water levels are low or medium, the motor rotation speed will be medium too.

1. Introduction

It is highly important for studying a Water Level Control (WLC) especially in the industrial implementation like a boiler and other liquid tanks in nuclear power plants [1]. Nowadays the use of electric energy increases. In 2013, the use of electric energy in Indonesia increased by 7.2% from 2012, i.e. from 84.43 Tera Watt Hour (TWh) to 90.48 Tera Watt Hour (TWh) [2]. This one was caused by the waste of home electricity use. Among the system's power consumption that is often used is the utilization of electric motor as water pumps in the tank. One of the equipment that uses an electric motor is the water pump that is normally controlled by a Water Level Control (WLC) [3]. It is necessary several techniques to achieve a safe and comfortable [4].

During the time, found high water control system using the float as a switch for disconnect and connect the water pumps, e.g. with mercury float switch [5]. Using conventional methods, to detect the level of liquid within a tank, generally using a float switch. Normally, the switch can be operated in an indicator, an alarm and a pump.

IOP Conf. Series: Materials Science and Engineering **535** (2019) 012021 doi:10.1088/1757-899X/535/1/012021

However, the WLC system is the uncertainties and nonlinearities that can be categorized as a complex system. Many methods have been used to obtain the WLC system is effective and efficient. In most residential areas are used combined water tank system.

Conventional control methods cannot support to make a convenient and solve the complexity [6]. It is necessary an intelligent control system to solve it. An alternative technique that can be implemented for developing a nonlinear and linear system using intelligent system [7]. Reshmi [1] presented Fuzzy Logic Controller (FLC) and implemented this technique using Matlab and Simulink to assess the WLC behaviour. Similarly, with the work to simulate of FLC for water tank level reached by [8], it was realised by combining between FLC and PID control and using simulation package that is Fuzzy Logic Toolbox and Simulink within MATLAB software.

Several methods can be used in developing an FLC technique, namely Mamdani, Sugeno and Tsukamoto. Dhanya et al [9] studied Mamdani method to assess the system for various inputs through the Fuzzy Inference System (FIS) Toolbox within Matlab for solving the uncertainties of WLC, the FLC can consider to solve an automation problem of WLC.

However, to make a comparison between a conventional control system and FLC in order to solve uncertainties problems of WLC, [10] explain comparison results between PID and FLC. The comparison conducted with kinds of methods namely integral squared error (ISE), integral absolute error (IAE), integral of time and squared error (ITSE) and integral of time and absolute error (ITAE) and the performance of FLC more enhanced compared to PID performance.

For the experimental works, [11] applied WLC through FLC within the hardware for the switching namely transistor and then the data processor using PIC16 microcontroller and then connected to a PC with serial communication.

This paper presents a system control WLC with FLC using Sugeno method for simulation and experimental works. Only a few studies that presented comparing between simulation and experimental works for WLC based FLC. This paper uses the input data with Light sensor FLC Dependent Resistors (LDR) as the outdoor light detector and ultrasonic sensors to detect high low water in the tank. While the output on a WLC with this FLC is the round motor that will pump water to be appointed into the tank so that it can fully automatically.

The results show that the lighter of the outdoors and the higher low water, the water-pumping motor speed is faster or maximal. While the high water is medium and the bright light or daylight then the motor rotation speed is medium and when the water is maximal then the motor will die. Conversely, if the outdoor light is darker or at night and water levels are low or medium, the motor rotation speed will be medium. The challenge of this paper is presented by comparing between simulation and experimental works for WLC based FLC. The result shows similar results as simulation results using Matlab and Proteus can be obtained by extensive effort in experimental works.

2. Method

The sensors applied in this study were LDR and ultrasonic of SRF04 that were assembled at the outdoor side and the inside of tank respectively. The ultrasonic sensor of SRF04 has four pins namely GND, VCC, ECHO and Trigger. Pins of Echo and Trigger were connected to the microcontroller with of pins of PINA.0 and PINA.1 respectively.

The input of analogue was raised both LDR and ultrasonic were processed through Analogue to Digital Converter (ADC) that within a microcontroller. Programming process in this study was conducted through the CPU of 2.90 GHz, Intel(R) Core™ i7 2.70 GHz, and RAM of 8.00 GB, the software and hardware capacity which were operated respectively. To monitor the system performance was showed using LCD monitor. However, the Microcontroller of Atmega 16 assembled for data processing in this study.

Control and data acquisition were accomplished by employing the microcontroller of Atmega 16 that provided a direct interface between the processor, motor and sensors through ADC for analogue input, analogue dan digital output. LDR and ultrasonic sensors will be driven by two analogue inputs were required in the experimental rig. The motor was used as representative for a Pompa air. The software required editor and compiler of AVR CodeVision, meanwhile it was necessary to edit and to compile. To assess the functionality and reliability of WLC system using LDR and ultrasonic sensors with the algorithm of FLC, it is necessary to confirm between water volume level and natural lighting from the sun.

WLC Automatic Control System is using FLC which consist of input and output. The number of inputs in this system there are 2 connected with 1 fruit sensors ultrasonic sensors and as well as the LDR output using either a round motor. Setting input LDR consists of two membership function (MF) that is day and night. While ultrasonic input consists of three MF i.e. low, medium and high. As for the motor output setting consists of three MF, namely off, medium, fast. So this study, there are 3 fuzzy variables, namely Ultrasonic Sensors, LDR and motor as water-pumping.

The LDR variable is divided into 2 sets of fuzzy, namely: night and day. While ultrasonic variable is divided into 3 sets of fuzzy, namely: low, medium and high. Meanwhile, variable motor divided into 3 sets of fuzzy, namely: off, medium and fast.

The universal set is a whole value that is allowed to be operated in a fuzzy variable. As for the fuzzy set domain variable on the LDR is:

Universal Set for LDR variable : [0 1000]
Universal Set for ultrasonic variable : [0 80]
Universal Set for motor variable : [0 255]

The domain of the entire set of fuzzy are values permitted in the Universal Set and can be operated in a fuzzy set. As for the fuzzy set domain variable on the LDR is

Night = [0:1000:1000]
Day = [0:0:1000]

While the set domain of fuzzy on ULTRASONIC variable is

Low = [0:0:40]
Medium = [0:40:80]
High = [40:80:80]

Meanwhile, the set domain variable of fuzzy on MOTOR is

Off = [0]
Medium = [127.5]
Maximal = [255]

The function of membership function (MF) is a curve that shows the mapping of input data points to engage in value that an interval between 0 to 1. While the output of automatic control system on the flame of water pump used the motor rotation.

The next step is the determination of the fuzzification on LDR input and ultrasonic. The specified input variables of LDR (x), namely the value of the analog to digital converter (ADC) for example with a value of 100. As for the determination of fuzzification with the MF Sugeno night model to get the error value (μ) is using the formula:

$$\mu\ MF\ Night = \begin{cases} 1; x \le a & \\ (b-x)/(b-a); & a \le x \le b \\ 0; & x \ge b \end{cases} \qquad (1)$$

Figures 1 and 2 show the membership function of ultrasonic input and motor as output respectively.

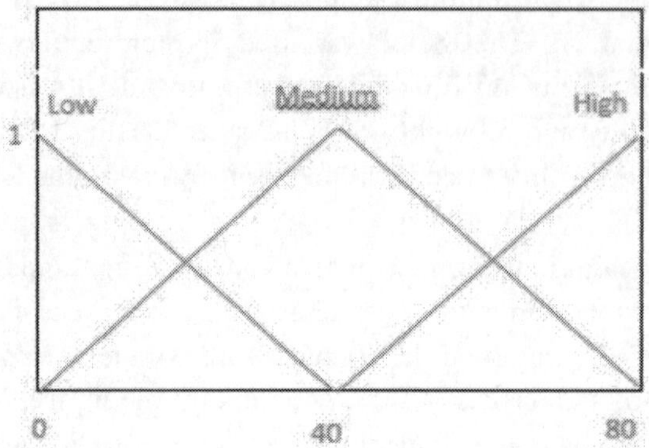

Figure 1. The membership function of ultrasonic input

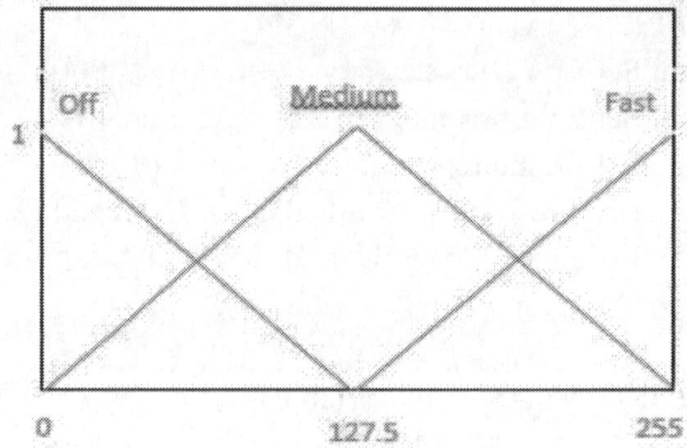

Figure 2. The membership function of output of motor

With x, b, a are input variable of LDR sensor, upper and lower bounds respectively. So, if for the night given MF x=100, a=0 and b=1000, it will get $\mu\ night = 0.9$

While the fuzzification determination with Sugeno model on the day MF to get the value of the error (μ) is by using the formula:

$$\mu\ MF\ Day = \begin{cases} 0; & x \le a \\ (x-a)/(b-a); & a \le x \le b \\ 1; & x \ge b \end{cases} \tag{2}$$

While x, b, a are input variable of LDR sensor, upper and lower bounds respectively, the MF day that is given x=100, a=0 and b=1000, will be got $\mu\ day = 0{,}1$.

For the type of the specified input variable ultrasonic input (x) is named the echo input and analog value output with a trigger to digital converter (ADC) for example with a value of 50. As for the determination of fuzzification with the Sugeno model on low to get the MF delta value error (μ) is using the formula:

$$\mu\ MF\ lower = \begin{cases} 1; x \le a & \\ (b-x)/(b-a); & a \le x \le b \\ 0; & x \ge b \end{cases} \tag{3}$$

As for the x, b, a for ultrasonic sensor are input variable of ultrasonic sensor, upper and lower bounds respectively. So, if MF day is given x=50, a=0 and b=40, will be got $\mu\ lower = 0$

As for the determination of fuzzification with the Sugeno model on medium to get the MF delta value error (μ) is using the formula:

$$\mu\ MF\ medium = \begin{cases} 0; & x \le a\ or\ x \ge c \\ (x-a)/(b-a); & a \le x \le b \\ (c-x)/(c-b); & b \le x \le c \end{cases} \tag{4}$$

As for the x, a, b, c for ultrasonic sensor are input variable of ultrasonic sensor, lower, medium and upper bounds respectively. So, if MF night is given x=50, a=0, b=40 and c=80, will be got $\mu\ medium = 0{,}75$.

While the fuzzification determination with Sugeno model on the high MF to get the value of the error (μ) is by using the formula:

$$\mu\ MF\ high = \begin{cases} 0; & x \le a \\ (x-a)/(b-a); & a \le x \le b \\ 1; & x \ge b \end{cases} \tag{5}$$

As for the MF high, x, b, a for ultrasonic sensor are input variable of ultrasonic sensor, upper and lower bounds respectively. So, if MF day is given x=50, a=0, and b=80, will be got $\mu\ high = 0{,}25$
Determination of RULES by using logic AND operation. $\alpha-predicate$ as a result of operations by using logic AND operation obtained with the smallest membership value between elements on the set correlation.

3. Results and Discussions

As for the RULE determination can be seen in Table 1 below.

Table 1. Rules of the FLC for WLC system

No	Rules
[R1]	If water LOW and LDR NOON then MOTOR FAST
[R2]	If water MEDIUM and LDR NOON then MOTOR MEDIUM
[R3]	If water HIGH and LDR NOON then MOTOR OFF
[R4]	If water LOW and LDR NIGHT then MOTOR MEDIUM
[R5]	If water MEDIUM and LDR NIGHT then MOTOR MEDIUM

[R6]	If water HIGH and LDR NIGHT then MOTOR OFF

The next step in determining rules on the preparation of FLC is the implementation of the rules in programming. Table 2 shows the equation process on implementation rules.

Table 2. Rules implementation of the FLC for WLC system

No	Rules
[R1]	$\alpha - predicate_1 = \mu ULTRAlow \cap LDRnoon$
[R2]	$\alpha - predicate_2 = \mu ULTRAmedium \cap LDRnoon$
[R3]	$\alpha - predicate_3 = \mu ULTRAhigh \cap LDRnoon$
[R4]	$\alpha - predicate_4 = \mu ULTRAlow \cap LDRnight$
[R5]	$\alpha - predicate_5 = \mu ULTRAmedium \cap LDRnight$
[R6]	$\alpha - predicate_6 = \mu ULTRAhight \cap LDRnight$

After the process of determining the rules finished then continued with the process of defuzyfikasi so that it can determine the value of the output (Z) using the following formula,

$$Z = \frac{\alpha pred_1 * z_1 + \alpha pred_2 * z_2 + \alpha pred_3 * z_3 + \alpha pred_4 * z_4 + \alpha pred_5 * z_5 + \alpha pred_6 * z_6}{\alpha pred_1 + \alpha pred_2 + \alpha pred_3 + \alpha pred_4 + \alpha pred_5 + \alpha pred_6} \quad (6)$$

So that the output value of the amount obtained (Z) of 90.3125 that will be used to set the pulse width of PWM on the motor. Meanwhile, after the completion of the design and simulation of FLC then continued with experimental works.

In the experimental works, the hardware component namely input needed the form of 1 piece sensor LDR and 1 fruit ultrasonic sensors as well as an electric motor as the output. Voltage source Vcc and ground on the sensor are mounted by using the socket resources (+) and (-).

Preparation of auto-control system of water pump using the FLC was designed using an application CV AVR. With block diagram can be seen in Figure 3.

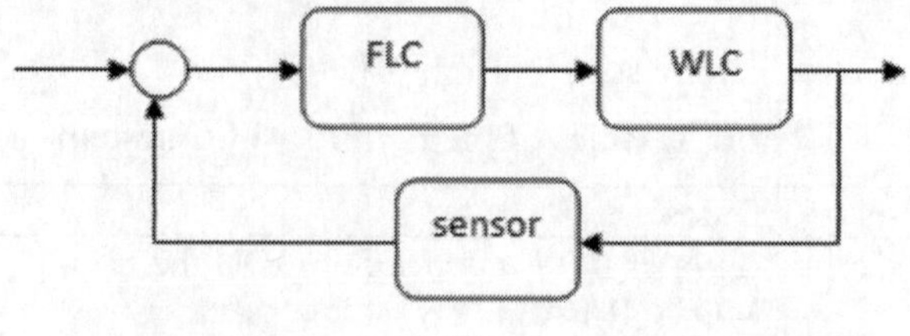

Figure 3. The schematic diagram of WLC

In this study has been conducted simulation and experimental works. In experimental work, the sensors are used namely SRF04 ultrasonic has 4 legs i.e., GND, VCC, ECHO and TRIGGER. ECHO is set as the input on the PIN ADC (0)/PINA. 0 and TRIGGER are set as the output on PIN ADC (1) or PINA. 1.

While the output that is a Motor Driver of L293D mounted in PIND. 0 and the PIND. 1 for input and Enable on PINB. 3. With the output, driver motor is connected to the Motor. It is necessary for connecting the ground terminal of the motor to the socket (s).

As for Pulse Width Modulation (PWM) used as Motor speed regulator. When the characters smaller than PWM, then PORTD.0=1 (motor output PIND.0 value 1 or active) and when over PWM, then PORTD.0=0 or motor is off.

Before the test is done on the hardware, the first performed testing using simulation software Matlab and Proteus. Figure 4 shows the electronics circuits of WLC using Proteus.

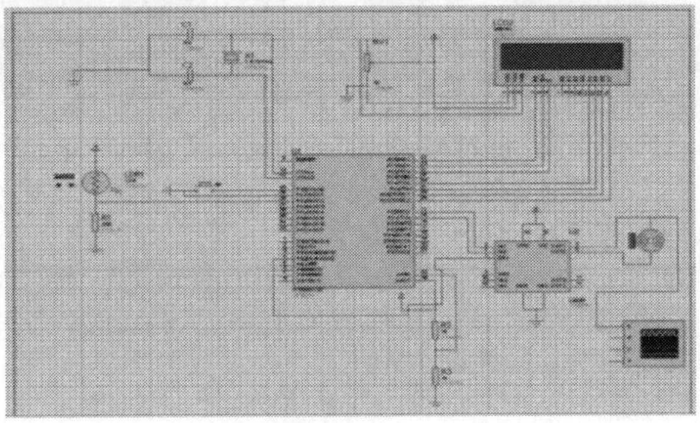

Figure 4. The electronics circuits of WLC

PWM used as a speed regulator Motor. When the characters smaller than PWM then PORTD.0=1 (output motor PIND.0 value 1 or active) and when over PWM then PORTD.0=0 or the motor is off. While the sensor ultrasonic used to measure three distance i.e. low, medium, and high. To find out the limitations, looked for μ with MIN-MAX method. LDR is used as a determinant of night and day. μ night or u night have 1 value if LDR 0 and have 0 value if LDR 1000. The performance measurement results on the experimental work results can be presented at Table 3.

Table 3. Experimental Performance of FLC for WLC

No.	INPUT		Output of PWM Motor			
	Ultrasonic (volt)	LDR (lux)	Experimental	Calculation	Simulation Using Matlab	Simulation Using Proteus
1	5	24	130	130,42	130	130
2	5	4	128	128,01	128	128
3	0	24	131	130,56	131	131
4	0	12	129	129,03	129	129
5	5	420	170	170,34	170	170
6	5	204	148	148,31	148	148
7	0	40	133	132,60	133	133

The results show that in the lighter outdoors and high water is low then water-pumping Motor speed is the faster or maximal. When the high water is medium and a bright light or daylight then the motor rotation speed is medium and the water is maximal then the motor will die. Conversely, if the outdoor light is darker or at night and the water levels are low or medium, Motor rotation speed will medium too.

Based on Table 3, it is noted that the comparison between the calculation results of the output with the results of experimental work is having similarity and only a very small error occurs. As for the error value between experimental results and calculation, works can be presented with Mean Square Error (MSE) of 0278. It is noted that performance of experimental work is closer with simulation work.

It is also shown with simulation work based on Matlab and Proteus, the performance of experimental work is similar to simulation work. On Table 3 also shows numerous different variations of outdoor light conditions and water levels that detected by ultrasonic and LDR respectively. The results show that although with different variations of the input value of LDR and ultrasonic then keep producing performances similarities between simulation and experimental works.

4. Conclusion

In this study has developed WLC system based on FLC with Sugeno technique. The FLC has input two input and single output namely LDR, ultrasonic and motor respectively. In this study presented simulation work within Matlab and Proteus and then validated through experimental work. The result shows that the simulation work is much closer to experimental work. Only a few errors between calculation method and experimental work. The FLC smoothly can the WLC. The FLC performance between experimental and simulation show similar.

5. References

[1] Reshmi Banerjee, "Water Level Controller By Fuzzy Logic," International Journal of Innovative Research in Advanced Engineering, Vol.2 No. 2, 2015

[2 M. Khairudin, S. Riyadi and Z. Mohammed, "Development of Fuzzy Logic Control for Indoor Lighting Using LEDs Group," Telkomnika, Vol. 16. No. 3, 2018.

[3] Asaad Ahmed MAE and Zhang JM, "Automatic Water Level Control System," International Journal of Science and Research,Vol. 4 No. 12, 2015.

[4] Ismail ME, Masran S.H, Rahim M.B, Faizal A.N and Marian M.F. "Development of Electrical Discharge Machine Die Sinking Application Using Android Platform," Jurnal Pendidikan Teknologi dan Kejuruan, Vol. 23, No.4, 2017

[5] Erua J. Band and Anyasi F. I, "Design of an Automatic Water Level Controller Using Mercury Float Switch," IOSR Journal of Electronics and Communication Engineering, Vol. 9, No. 2, 2014: PP 16-21.

[6] Khairudin M. "Quantitative Feedback Theory-Based Robust Control For A Spindle Of Lathe Machine" International Journal on Smart Sensing and Intelligent Systems, Vol.9 No. 4, December 2016.

[7] Febri SR, Fatchul A. "Developing Artificial Neural Network Based on Visual Studio for Dance Assessment," Jurnal Pendidikan Teknologi dan Kejuruan, Vol. 23, No.4, 2017

[8] Dharamniwas, Aziz Ahmad, Varun Redhu and Umesh Gupta, "Liquid Level Control By Using Fuzzy Logic Controller," International Journal of Advances in Engineering & Technology, Vol. 4, No. 1, 2012: 537-549

[9] Dhanya S, Irene George, Giffy Francis, Karthika R and Bobby Abraham, "Constant Water Level Controller Using Fuzzy Logic," IJARIIE, Vol.2, No.3 2016: 111-121.

[10 Laith Abed Sabri and Hussein Ahmed, "Implementation of Fuzzy and PID Controller to Water Level System using LabView," International Journal of Computer Applications, Vol. 116, No. 11, 2015: 6-10

[11] Namrata Dey, Ria Manda and M Monica Subashini, "Design and Implementation of a Water Level Controller using Fuzzy Logic," International Journal of Engineering and Technology, Vol 5 No 3, 2013: 2277-2285.

IOP Conf. Series: Materials Science and Engineering **535** (2019) 012022 doi:10.1088/1757-899X/535/1/012022

The evaluation of industrial assisted classes program in vocational schools in Yogyakarta

S Hadi[1] and Rabiman[2]

[1,2]Department Of Mechanical Engineering Education, Sarjanawiyata Tamansiswa University, Jl. Batikan, Tuntungan UH III/1043 Umbulharjo, Yogyakarta, Indonesia

E-mail: sam.otto@yahoo.com

Abstract. This research was aimed to know the attainment of industrial assisted classes program in vocational school (SMK) in the special region of Yogyakarta. This research is an evaluation model using CIPP adopted from Stufflebeam. The setting of the research is SMK in special region of Yogyakarta (DIY) which organize the program of industry class which are: SMK Muhammadiyah 1 Imogiri Bantul, SMK Muhammadiyah 1 Playen Gunungkidul, SMK Muhammadiyah Pakem Sleman, SMK Muhammadiyah 3 Yogyakarta and PT. AHM Yogyakarta. Informants in this study are the SMK Manager, Productive Teacher Motorcycle Engineering, and Honda's Training Manager. The collection of data through observation, in-depth interviews, and documentation. Data analysis uses an interactive model from Miles & Huberman. The results of the study concluded that: (1) the evaluation of contex: SMK and industry have agreed that the minimum graduates at level I, SMK-Industri have both understood the policy and objectives of the built-in class program, SMK has prepared the standardized provisions of the industry; (2) the evaluation of input: the qualification and competence of productive teachers are in accordance with the minimum requirement of S1 and have minimum training certificate of brown level, the school curriculum has been integrated with the industry curriculum, adjustments of the learning facilities and infrastructure to industry standards is done gradually, the schedule of learning is the authority of schools and industry assist scheduling for field work program and industry visit program; (3) the evaluation of process: the industry help the preparation of learning materials, training, the provision of places of visit and fieldwork practices of industry, industry monitoring and supervision, the industry conduct the test of competency; (4) the evaluation of the product: 60% of the industry workforce partners come from the industrial assisted classes program graduates, the granting of aid to students who are going to entrepreneurship.

1. Introduction

The alignment of the education world as a supplier of human resources with a working world that has dynamic needs and demands is a necessity. The concept of alignment implies the need for good coordination between the providers of education graduates and those who need graduates. Optimal alignment efforts through the implementation of a systematic and continuous series of programs between education and the industri are essential. An example of what is happening today is the collaboration between SMK with business and industry, such as Dual System Education (Internship, PKL, Prakerin) and establishment of The industrial assisted classes program.

The industrial assisted classes are the result of the collaboration undertaken by the SMK and Industri in filtering the seeds of students who excel in the school to be educated in accordance with the planned curriculum industri in one class. All facilities and infrastructure at least close to the replica of the industry. The concept adheres to Prosser C A & Quigley T H (1950) suggested that, "Vocational education will be efficient in proportion as the environment in which the learner is trained is a replica of the environment in which he must subsequently work"[1]. The existing vocational education within industrial assisted classes will be effective if it is in harmony with the existing environment in the industry.

Growth in the establishment of the industrial assisted classes program demands on the success of all components on evaluation. During this time, the evaluation has not been fully done by schools or industry. Implementation of the evaluation is limited to the implementation process in the classroom. Comprehensive evaluation has not been carried out completely, so that the weakness/obstacle in each component, such as: context, input, process, and product have not been detected deeply. The evaluation leads to (1) Context evaluation, that covering: (a) the quality of student competence, (b) policies and objectives of the industrial assisted classes program, and (c) preparedness in the management of the industrial assisted classes program; (2) input evaluation, including: (a) human resources, (b) curriculum, (c) learning support facilities, and (d) learning time / schedule; (3) process evaluation, including: (a) learning planning, (b) implementation of learning, (c) monitoring and supervision of learning, and (d) maintenance and maintenance of facilities and infrastructure; and (4) product evaluation, covering: (a) the quality of graduate students, (b) the absorption of graduates in the world of work, and (c) increasing the number of students

This evaluation aims at knowing during the process whether the goal of education has been achieved (Tyler in Suharsimi & Safrudin, 2009)[2]. The purpose of a vocational education that fully produces skilled, competent, creative, innovative, and active graduates should be able to reflect what industry needs. There are many cases in the industry that cause large losses to the industry due to human error, unemployment in work, and not in accordance with the qualifications required. Program evaluation through Stufflebeam's CIPP (Context, Input, Process, Product) model is expected to know the constraints / obstacles, assist administration in making decisions, and to describe, and provide information related to overall system improvement. The results of the evaluation of this program are expected to be used as a reference for other SMK to open industry-targeted classes.

Vocational education has been known for a long time and proclaimed by Master Vocational Education, Prosser C A. Prosser C A & Quigley T H (1950) formulated 16 propositions for vocational education as follows: (1) "Vocational education will be efficient in proportion as the environment in which the learner is trained is a replica of the environment in which he must subsequently work (2)"Effective vocational training can only be given where the training jobs are carried on in the same way with the same operations, the same tools and the same machines as in the occupation itself (3) Vocational education will be effective in proportion as it trains the individual directly and specifically in the thinking habits and the manipulative habits required in the occupation itself (4) Vocational education will be effective in proportion as it enables each individual to capitalize his interest, aptitudes and intrinsic intelligence to the highest possible degree (5) Effective vocational education for any profession, calling, trade, occupation or job can only be given to the selected group of individuals who need it, want it, and are able to profit by it (6) Vocational training will be effective in proportion as the specific training experiences for forming right habits of doing and thinking are repeated to the point the habits developed are those of the finished skills necessary for gainful employment (7) Vocational education will be effective in proportional as the instructor has had successful experience in the application of skills and knowledge to the operations and processes he undertakes to teach (8) For every occupation there is a minimum of productive ability which an individual must possess in order to secure or retain employment in that occupation. If vocational education is not carried to that point with that individual, it is neither personally or socially effective (9) Vocational education must recognize conditions as they are and must train individuals to meet the demands of the "market" even though it may be true that more efficient ways of conducting the occupation may be known and that better working conditions are highly desirable (10) The effective establishment of process habits in any learner will be secured in proportion as the training is given on actual jobs and not on exercises or pseudo jobs (11) The only reliable source of content for specific training is an occupation is in the experience of masters of that occupation (12) For every occupation there is a body of content which is peculiar to that occupation and to which has practically no functional value in any other occupation (13) Vocational education will render efficient social service in proportion as it meets the specific training needs of any group at the time that they need it and in such a way they can most effectively profit by the instruction (14) Vocational education will be socially efficient in proportion as in its methods of instruction and its personal relations with learners it takes into consideration the particular characteristics of any particular group which it serves (15) The administration of vocational education will be efficient in proportion as it is elastic and fluid rather than rigid and standardized (16) While every reasonable effort should be made to reduce per capita cost, there is a minimum below which effective vocational education cannot be given, and if the course does not permit this minimum per capita cost, vocational education should not be attempted[1].

Based on the theory described by Prosser & Quigley, it can be concluded in general that the implementation of vocational education in SMK will be effective if the entire environment, culture, conditions, human resources, and infrastructure must replicate / replicate with industry. Thus vocational education will be effective if it can train students directly and specifically in habits of thinking and working in accordance with the type of work[1].

The industrial development class is the cooperation class between Vocational High School and industry. The class is specially designed to produce graduates in accordance with the industry's expected qualifications. In accordance with Prosser & Quigley (1950) statement which explains in one of his theories, "The training environment is the working environment or a replica of the working environment." An industry-appropriate learning environment can shape thinking habits, work, and behave, so that students are better prepared and confident in facing the challenges encountered while working[1].

PT Astra Honda Motor (AHM) is one of the industries that cooperates with Vocational High School (SMK) in organizing industry-class programs. The number of PT AHM target classes throughout Indonesia until 2017 ± 670 units. Implementation of e-learning in some vocational schools in Bantul district is still very poor, so very few students can use it. Some vocational schools that implement e-learning can use it as a medium of learning in school. Utilization of e-learning is expected to improve competence and support adaptive learning of SMK students in Bantul District.

Program evaluation is a series of deliberate activities to see the success of the program. Understanding the program according to Big Indonesian Dictionary is a plan or activities done carefully. According Suharsimi & Safrudin (2009), program evaluation aims to find out how high the success rate of planned activities. The evaluation model of the program that is widely applied in the evaluation research is the CIPP Model (Context, Input, Process, and Product). The following is an explanation of each evaluation component[2].

1.1. Context Evaluation

Context evaluation is the first evaluation required in the CIPP model. An explains that the main orientation of the context evaluation is to identify the background of the need to make changes or the emergence of programs from some of the subjects involved in decision making. Suharsimi & Safrudin (2009) added that context evaluation is used to answer questions: (a) what needs have not been met by program activities; (b) which development objectives are related to the fulfillment of needs; and (c) which goals are the easiest to achieve. These questions will generate inputs to make alternative decisions regarding plans and strategies for implementing future programs[2].

The evaluation of the context in relation to this research refers to several aspects: (a) the quality of student competence, (b) policies and objectives of the The industrial assisted classes program, and (c) preparedness in the management of The industrial assisted classes program. By definition, the quality of students' competence is the quality that students possess, including knowledge, skills, and behaviors that are lived and actualized in a skill area. Initial quality of the students needs to be evaluated in order to map how much quality the student has in preparing an industry-class opening plan.

Another aspect that needs to be evaluated is the policies and objectives of the industrial assisted classes. Policies and objectives of industry-led classes according to Tomoki Uchida one of them is "preparing human resources in Indonesia who are ready to work and able to compete with resources from other countries, especially at Southeast Asia Regional level" (Metrotvnews.com, March 16, 2016)[3]. This objective will be aligned with the the Indonesian education goals stated in Undang-Undang Sisdiknas No. 20 in 2003. It states that education is a conscious and planned efforts to achieve a good teaching and learning process so that the learners are able to develop their potential capacities actively for gaining spiritual, self- control, personal, intellectual strengths as well as appropriate skills needed for themselves, their society and their country[4].

The next aspect that needs to be evaluated is the readiness of schools in managing industry-led classes. School readiness in managing a quality program can refer to the National Standards of Education.

1.2. Input Evaluation

The definition of input according to Zaenal Arifin (2011) is the process and efficiency of resources, tools, people and costs, to implement the programs that have been selected. The main purpose of input is to determine how to use inputs to achieve program objectives. Feedback is very diverse, ranging from coaching programs, financing, obstacles, selected strategies, learning facilities, the environment, facilities and infrastructure, quality of human resources, and so forth. Based on Prosser's theory, the aspects that need to be evaluated especially in the understanding of industrial grade can be seen from: (a) human resources such as teachers, instructors, technicians, and mechanics, (b) curriculum industry, (c) learning support facilities, (d) preparation of funds and funds, and (e) time / schedule of learning[5].

Human resources (HR) that includes teachers, instructors, and mechanics need to be evaluated to synchronize with real conditions in the industry. Teacher-specific teacher qualifications should be lit on 4 items under Law No. 14 in 2005 on Teachers and Lecturers covering: pedagogic competence, personality competence, social competence, and professional personnel. These four goals must be mastered by teachers and actualized to the students.[6]

The next aspect that needs to be evaluated is the industrial feed. The curriculum that develops in vocational schools still chooses on a competency-based curriculum. The curriculum still contains some points that are not yet compatible with the industry's required curriculum. This evaluation aims to show the relevance of SMK curriculum to industry.

Further input input is a means of supporting learning both facilities and infrastructure. Etymologically, means and infrastructure means a direct and indirect means of achieving educational goals. Facilities include: Space, Books, Libraries, Laboratories, etc., while infrastructure includes: location / place, school building, sports field, money, and so on.

This evaluation was done to find out the consistency of facilities in industry. The next evaluation is budget and fund preparation. The evaluation leads to school revenues and expenditures. Benefits can be sourced from School Operational Assistance (BOS), school committees, foundations, production units, and other sponsors, while other programs such as: fixed expenditure, non-fixed expenditures, isidental expenditures, vaccations, etc. that need to be tailored to the minimum needs must exist in accordance with the conditions in the industry.

1.3. Process Evaluation

Process evaluation aims to identify or predict obstacles in the implementation of activities or program implementation. According to Zaenal Arifin (2011), evaluation is done by recording or documenting every event in the execution of activities, monitoring activities that potentially hamper and causing unexpected difficulties, finding out special information that is beyond the plan; assessing and explaining the process in actuality[5].

During the evaluation process, evaluators are required to interact with the program implementing staff on an ongoing basis. Components that need to be evaluated include: (a) learning planning, (b) implementation of learning, (c) monitoring of learning, and (d) maintenance and maintenance of facilities and infrastructure.

Learning planning component needs to be evaluated process to know readiness and planning of teacher. The lesson plans that have been contained in the RPP and Syllabus must be properly prepared by the teacher / instructor. In the RPP and Syllabus teachers need to prepare and design what kind of Competency Standards, Basic Competencies, Character Values, Learning Materials, Learning Activities, Activity Type, KKM, Assessment, Media, Methods, Time Allocation, and Learning Resources to be used in accordance with direction of industry. The next component is the implementation of learning. Evaluation of the process in the implementation of learning only implements what has been poured in RPP, classroom management, delivery of materials, interaction and communication, and guidance learning.

1.4. Product Evaluation

Product evaluation is the evaluation of output (output). Evaluation of directional outputs on direct results (direct) program. Both changes to the teaching performance of the educator as well as the student learning performance observed at the end of the program implementation. The results of this targeted classroom program are the student competencies expected by the industry. Product evaluation components that can be evaluated include: (a) the quality of students, (b) the absorption of graduates in the world of work, and (c) the increase in the number of students.

The quality of students can be seen in the competence of students who have. Student competencies covering aspects of attitude, knowledge, and skill can be measured by competency test. The industry's expected qualification is industry-standardized qualifications for students. The quality of students can affect the graduates' absorption of the cooperating industry. The impact that schools can feel is an increase in the amount of interest.

Research questions that should get answers in this study are: how the achievement of the implementation of the industrial assisted classes program reviewed in context, input, process, and product evaluation.

2. Method

This research is an evaluation research with the approach used is qualitative. The evaluation evaluation model used adopts CIPP from Stufflebeam which includes context, input, process, and product. This research uses several steps as follows: (1) Preliminary surveys to industries that organize industry-class programs. Survey results will be used as guidance in conducting process and examination instruments and questionnaires in schools (2) Preparation of instruments starting from lattice instruments, interviews, expert test instruments, and limited trials. The grid compiles based on data obtained from the industry (3) Data collection in industries and schools, both on context, input, process and product components (4) Analysis of data on each component context, input, process, and product (5) Reporting of research results.

This research was conducted on 4 (four) SMK in DIY which implement industrial-grade program and PT. Astra Honda Motor (AHM) Yogyakarta. Research subjects are divided into two places. The first research subject is Training Manager of PT. AHM and the second research subject are the parties involved in the program, among others: principal, head of department, and productive teacher of Motorcycle Engineering. Data collection techniques in the evaluation research using several techniques, among others: observation, interviews, checklist and documentation. Instruments used include: checklist, interview guide, and documentation. Data analysis was conducted in two categories, namely: data analysis in the field to see the development of e-learning as a learning media and application of e-learning in school. The model of this research is continuous information digging, so that the resulting data can be met well and to produce saturation data. This data analysis uses an interactive model from Miles M B, Huberman A M, and Saldaña J. as Figure 1.

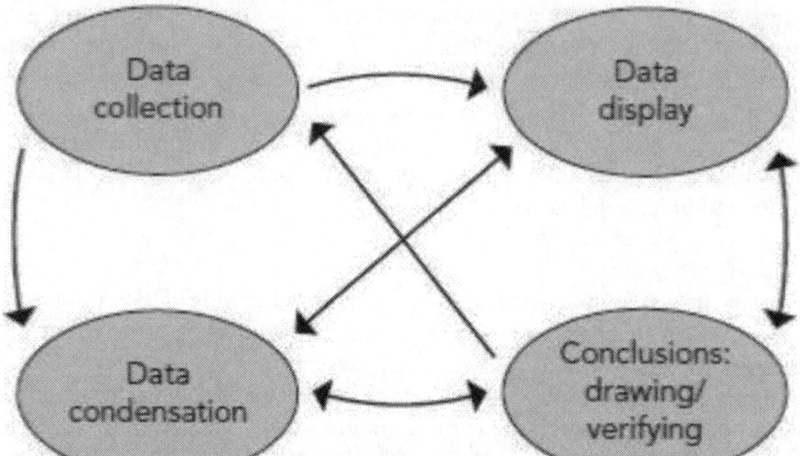

Figure 1. Data analysis components: Interactive Model (Source: Miles M B, Huberman A M, and Saldaña J, 2014: 10) [7]

3. Result and Discussion

The number of industrial-assisted classes in DIY until 2017 amounts to ± 13 units spread over 5 (five) districts. Industry expects the quality of graduates of industrial grade students at least have competence in level I, have a good attitude and personality. This has already become an agreement between the industry and SMK. This minimum level I competency becomes a reference in the implementation of The industrial assisted classes program. Industry and SMK already have the same views related to the policies and objectives of the organization of The industrial assisted classes program. The industrial assisted classes program are organized in order to provide graduates to work in partner industries under PT. AHM and / or entrepreneur in the field of motorcycle maintenance and repair especially honda.

SMK has been providing human resources, especially productive teachers, learning infrastructure tools as standardized by industry such as (layout / layout of learning place, area, practice tools, and practice objects) and preparing curriculum adapted to industry-specific curriculum. Productive Teacher Motorcycle Engineering is preferably S1-educated and has graduated in instructor training held at a minimum level of brown industry. Industry provides competency training to teachers to obtain training certificates (Brown, Silver, and Gold).

The school curriculum has been integrated with the industry curriculum. The use of curriculum from industry is very helpful in curriculum development in schools. However, for the adjustment takes approximately 1 (one) year. The parties involved in the adjustment of the curriculum include industry, productive teachers and related education offices. The alignment of school learning facilities with existing facilities in the industry is done gradually adjusting the ability and readiness of schools. The determination of the learning schedule is entirely left to SMK, the industry helps the scheduling of internship / field work practices and industry visits.

The learning process of built-in classes is replicated as it is in the industry. Industry helps in the learning process especially on the preparation of teaching materials (module, learning evaluation, practice object, information technology development and procurement competition). Industries assist skills training to productive classroom teachers. The industry has developed a special E-learning system for the built classes that can be accessed by the built-in SMK. With this E-Learning system SMK assisted communication with industry can run well and quickly. In addition the industry provides venues for internships / fieldwork practices as well as places for well-scheduled industry visits.

Industry always conduct monitoring and supervision to SMK-SMK targeted either directly by visiting SMK or with E-Learning system. Industry conducts monitoring and supervision 4 (four) times a year. The industry has also undertaken an assessment to rank for the assisted SMK (A, B, and C) and simultaneously mapping this will facilitate the industry in conducting development and development.

The industry provides assistance in the addition of learning infrastructure facilities. Assistance provided to SMK is tailored to the needs in the field. While the maintenance and repair performed by SMK.

Industry together with the National Accreditation Body (BAN) SMK held a learning evaluation, especially for student competency test. The industry has prepared the human resources of assessors team to conduct student competency test. Industry participates in conducting student competency tests along with school assesors to maintain the quality of graduates. The industry is responsible for the graduation rate of graduates that is with the policy that 60% of workforce in partner industry of PT. AHM must come from SMK target. With this policy, it is expected that graduates of the assisted class can be well absorbed. Industry helped promote SMK. With this class, SMK promotion is better. This is proven by the increase of student interest to enter the program of motorcycle engineering expertise of the built classes.

4. Conclusion

Based on the results of research and discussion of this study, it can be drawn conclusion as follows: (1) Competency Standards Graduates of grade students are minimal at Level I level. This standard is used as the basis in the learning process of the target classes (2) Industrial policies and objectives in the implementation of the target classes are in accordance with the school's policies and objectives (3) School readiness in the management of The industrial assisted classes program is done gradually by adjusting school readiness and needs (4) Competence and qualification of productive teachers of The industrial assisted classes program are in conformity with standards determined by the industry. Industry is responsible for continuous improvement of teacher skill competence (5) The school curriculum has been integrated with the industry curriculum (6) Alignment of school learning facilities with existing facilities in the industry is done in stages (7) Schedule of learning is determined fully SMK, industry helps scheduling for internship / field work practice and visit to industry (8) With Industrial E-learning system helps in the learning process especially on the preparation of teaching materials (module, learning evaluation, practice object, giving information of technological developments and competitions (9) Industry helps to monitor and supervise to SMK either directly to SMK or E-Learning system and give rating to SMK (Rating A, B, and C) (10) The industry is committed to providing learning infrastructure facilities. While the maintenance and repair performed by SMK (11) Industry together with BAN SMK held a learning evaluation especially for student competency test. The industry has prepared accesor resources to conduct student competency tests (12) Industry is committed to participate in maintaining the quality of graduates (13) The industry is responsible for the absorption rate of graduates by applying the policy that 60% of labor in partner industry of PT. AHM must come from SMK target (14) The interest of new students continues to increase with the presence of the target classes

5. References

[1] Prosser C A and Quigley T H. (1950) *Vocational Education in a Democracy*. Revised Edition. (Chicago: American Technical Society)

[2] Suharsimi *and* Safrudin C 2009 *Evaluasi Program Pendidikan* (Jakarta: Bumi Aksara)

[3] Metrotvnews.com 2016 *Honda Mobil Aktif Bangun Dunia Pendidikan Nasional*. *http://otomotif.metrotvnews.com*. 16 Maret 2016

[4] Depdiknas 2003 Undang-Undang Republik Indonesia No. 20 Tahun 2003 Tentang Sistem Pendidikan Nasional.

[5] Zaenal Arifin 2011 *Evaluasi Pembelajaran: Prinsip, Teknik, dan Prosedur* (Bandung: PT. Remaja Rosdakarya)

[6] Depdiknas 2005 Undang-Undang Nomor 14 Tahun 2005 Tentang Guru dan Dosen

[7] Miles M B, Huberman A M, and Saldana J 2014. *Qualitative Data Analysis, A Methods Sourcebook*, Edition 3. (London: Sage Publications)

IOP Conf. Series: Materials Science and Engineering **535** (2019) 012023 doi:10.1088/1757-899X/535/1/012023

Tracking the potential of vocational high school students leading to eighteen strategic creative industry sub-sectors in Indonesia

BR Setiadi[1], Suparmin[2], S Priyanto[3] and P Roniwijaya[4]

[1,2,3,4]Sarnawijaya Tamansiswa University, Yogyakarta, Indonesia

E-mail: bayurahmatsetiadi@ustjogja.ac.id

Abstract. This study aims to trace the interest of Vocational High School (VHS) students to meet the sub-sectors of strategic creative industry in Indonesia which is divided into eighteen sub-sectors such as animation, architecture, design, photography, music, crafts, culinary, fashion, research and development, publishing, film, advertising, interactive games, performing arts, visual arts, information technology, television, radio and video. The study will also track the reasons why students choose these sub-sectors with three priority-scale options.The research method used is an exploratory survey with the dominant approach is quantitative. This study was conducted in 7 VHS in DIY-Indonesia with a sample size of 350 people. The subjects of this study are students with a background of educational concentration is technology and engineering. The technique of collecting data using questionnaire of specialization. Data analysis techniques used descriptive statistics that aims to determine quantitatively the percentage of student's interest in sub-sub creative industries.Quite unique findings are generated in this study because researchers want to explore the other side of the engineering students who in fact work by relying on hard skill instead of developing their soft skill. The results of this study indicate that the dominance of student interest to eighteen creative industry sub-sectors exists in the sectors of music, photography, and design. While the sub-sector that they are not interested in the fashion sector. Other tracking results are the reason students choose the three priority sectors of interest are because their hobbies are relevant to the sub-sectors of the creative industry of their choice. This study will have implications for schools to attend the potential of other gifted students in the creative industries. If this can be effectively explored, employment opportunities for engineering students will become wider and foster an entrepreneurial spirit so they can open new jobs for others in need.

1. Introduction

1.1 Problem of Statement

he creative industry has an important role in the development of the creative economy in developed countries. The impact of rapid progress in developed countries has been felt by developing countries, especially in increasing the country's foreign exchange. Data generated by United Nations Year 2003 states that 50% of G7 (United States, UK, France, Canada, Italy and France) come from creative industry products [1]. Indonesia as a part of developing countries in the world began to look at the creation of millions of employment since by compiling "The Development Plan of 2009-2015 Creative Economic".[1]

The existence of a policy of the development of creative industries, it should be directed to creative pilot cities in Indonesia such as Bandung, Yogyakarta, Surakarta, and Pekalongan [2]. The four big cities are a tourist destination that can provide satisfaction in cultural, nature, religious, entertainment and culinary travel based on Creative City Network (CCN) program developed by UNESCO in 2004. Therefore, the four creative cities become the focus of creative industry development in Indonesia considering that the inputs of these cities are very potential to be explored more comprehensively [3].

Daerah Istimewa Yogyakarta (DIY) is one of the provinces in Indonesia that is capable of producing young entrepreneurs capable of competing in regional, national and international scale. According to Vice Chairman of Indonesian Chamber of Commerce of Banking and Finance, the value of creative industry export in Jogja continues to increase, from the US $ 220 million in 2013 to the US $ 242 million in 2014 [4]. In fact, the Yogyakarta government is ready to disburse funds of 10 million rupiahs for each business group. This means DIY readiness in supporting the creative city of the world has been well responded by the government as a companion, director and facilitator for the creative industry in DIY can compete regionally, nationally and internationally.

Department of Industry, Trade, Cooperatives and Small-Medium Enterprises DIY noted that every year the number of creative industry businesses from 2012 to 2015 jumped 6-10% [5]. Fluctuations in these figures because of many specific sub-sectors such as handicrafts, fashion, and design that dominates almost all regions in DIY [6]. The growth of the three sub-sectors of the creative industries are less feel complete if it does not fill the sub-sectors other consisting of animation, architecture, design, photography, music, crafts, culinary, fashion, research and development, publishing, film, advertising, interactive games, performing arts, visual arts, information technology, television, radio, and video. The complete sub-sector of industry demands the role of the Indonesian government to encourage the entrepreneurial spirit of young people to transfer their capabilities to other creative industries sub-sectors in order to be equitable and strategic.

The fulfillment of the workforce in the creative industry varies greatly in educational background. Their driving motor is creativity [1] and brilliant ideas generated by business actors [7]. The results of field observations in some 20 industry actors in the sub-sector of creative industries in DIY showed that 55% came from SMK graduates and 45% came from high school graduates. This phenomenon shows that many SMK graduates who work in the creative industry even those who turn around work in the creative industry with a background of non-linear educational concentration. The transformation of SMK graduates who switched jobs into the focus of attention is important to explore why they prefer to work in the sub-sectors of the creative industry.

The direction of the vocational graduates' work can be traced back to the beginning by tracing the talents, interests, and potentials of students since class X. Contrast, talent, interest, and potentially developed in schools are more inclined toward discipline, work, technology, sports, and cultivation. Meanwhile, the expected work behind these fields is so small that it is profound that it produces high economic value. It is also no relevant to the Indonesian government's policy to accelerate millions of entrepreneurs in the creative industry sectors. The existence of this research will provide an overview of the vocational school student's interest, especially in preparing his graduate candidates to work in the creative industry labor market. Early detection provides inputs for schools especially in preparing graduates through self-development, extracurricular, as well as local content learning tailored to the growing creative industry sub-sector in DIY.

1.2 Purpose of Study

The existing research trends in Indonesia about the creative industry are much more studied about the world of work in the creative industry sector, while inputs from qualified human resources are rarely the concern of researchers in Indonesia. Human resource input can be obtained by finding talented talents to fill the sub-sectors of the creative industries so that there is equality of work sectors related to the development of local creative industries. Related to that, it is very appropriate with the purpose of the fulfillment of creative industry workers if we see the main vision of SMK is work, continue to college, and entrepreneurship.

This research will focus on mapping the interest of students of SMK X class in filling the sub-sector of creative industry in Indonesia in general and DIY in particular. The purpose of this study is to provide an overview of student's interest in 18 sub-sectors of creative industries in Indonesia as well as trace factors that influence students in selecting sub-sectors of the creative industry based on what they experience or interest.

2. Method

2.1 Research Design

The research used exploratory survey method with quantitative approach. The research will explore students' interest in the creative industry sub-sector in DIY based on their perception. Student perspectives can be a trend for the creative industries they are interested in and inform stakeholders to align the needs of the newest world of work. In the instrument is divided into two types namely the interest of students to each sub-sector and 3 alternative choices of sub-sector of the most interested by students. This research is part of the initial study of the potential empowerment model of vocational students for the development of creative industries in DIY. This model will further inform the direction and development of appropriate local creative industries to align with local school needs.

2.2 Research Sample

Research participants who contributed to this research are the students of class X of SMK in the field of technology and engineering expertise spread in 7 schools in DIY. The number of respondents is 350 students with the division in each school is 50 students so that the total class that is caught is 14 classes. The study program involved 3 classes of light vehicle engineering, 4 engineering classes, 1 class of computer and network engineering, 2 classes of building drawing techniques, 2 welding engineering classes, 1 mining engineering class, and 2 motorcycle engineering classes. This sampling is chosen randomly so that the researcher avoid the bias and subjectivity of research sample selection.

2.3 Data Collecting

Data collection in this study using questionnaires. The questionnaire used to collect data of student's interest in all sub-sectors of creative industries with 4 criteria Likert scale: very interest, interest enough, less interest, and not interest. This instrument consists of 32 statements that can represent each sub-sector of the creative industry and 3 alternative choices of student interest based on the highest level of interest preferred by the research respondents. The content of the questionnaire instrument is validated by an expert and declared eligible for a trial. The result of the instrument test with product moment correlation stated that the item 87.5% of the questions stated valid with the instrument reliability score is 0.838 based on the test with Alfa Cronbach's. After the validity and reliability requirements are met then the instrument is feasible to be used as data collection in the research sample.

2.4 Data Analysis

Data analysis used in this research use descriptive statistic. This is similar to the type of research used is a survey research method where researchers will explore the potential and interest of students in filling sub-creative industry sector in DIY. Analysis of this data is the result of student perceptions are poured in the questionnaire instrument based on 4 criteria of choice answers. This research will provide data and the quality of each student's interest to the creative industry sub-sector.

3. Result

The study was conducted in 7 vocational schools that are public and private. There are 4 public schools namely SVHS 3 Yogyakarta Regency, SVHS 2 Pengasih-KulonProgo Regency, SVHS 1 Sedayu-Bantul Regency, and SVHS 2 Depok-Sleman Regency. While private schools representing amounted to 3 namely PVHS National Berbah-Sleman Regency, PVHS Muhammadiyah 1 Bantul Regency, and PVHS 45 Wonosari- Gunungkidul Regency. The background of the study program is 14% motorcycle technique, 7% mining technique, 14% welding technique, 7% computer network engineering, 29% mechanical engineering, and 29% light vehicle technique. Based on this information shows that the areas of technology and engineering expertise dominated in this research are automotive engineering and mechanical engineering. The results of the questionnaire respondents tabulation showed 92% of respondents gendered men, and 8% female gender. The male-dominated tendency in this study is in line with the VHS competence map (DVHSD, 2017) recorded in the Vocational The Big Data's of SMK that the field of technology and engineering that many dominated by men [8].

The prospective creative industry as a focus of future work by students can be identified through the distribution of inquiry questionnaire. The questionnaire was first tested feasibility and consistency so deserve to be disseminated. The results obtained from each school can be reflected in the interests of the sub-sectors of creative industries by referring to the following table:

Table 1. Specialization of creative industry sub-sectors based on students' perception

No.	Creative Industries Sub-sector	School name							Mean
		SVHS 3 Yk	SVHS 2 Dpk	SVHS 2 Pgs	SVHS 1 Sdy	PVHS 1 Bnt	PVHS Nas	PVHS 45 Wns	
1	Animation	0,62	0,50	0,56	0,58	0,50	0,60	0,56	0,56
2	Architecture	0,60	0,52	0,58	0,68	0,56	0,57	0,65	0,59
3	Design	0,67	0,66	0,58	0,63	0,57	0,61	0,72	0,63
4	Photography	0,70	0,60	0,67	0,62	0,69	0,69	0,65	0,66
5	Music	0,65	0,61	0,64	0,66	0,66	0,67	0,70	0,66
6	Handycrafts	0,65	0,61	0,55	0,66	0,58	0,61	0,62	0,61
7	Culinary	0,65	0,66	0,58	0,60	0,55	0,55	0,65	0,61
8	Mode	0,53	0,50	0,49	0,45	0,44	0,44	0,48	0,48
9	Research & Development	0,50	0,59	0,57	0,46	0,51	0,52	0,51	0,52
10	Publishing	0,50	0,50	0,51	0,48	0,44	0,47	0,50	0,49
11	Film	0,60	0,48	0,56	0,51	0,46	0,55	0,54	0,53
12	Advertising	0,57	0,55	0,48	0,48	0,48	0,51	0,54	0,52
13	Interactive Games	0,67	0,51	0,55	0,60	0,51	0,55	0,56	0,56

14	Performing Arts	0,60	0,55	0,52	0,55	0,58	0,63	0,54	0,57
15	Arts	0,68	0,57	0,55	0,60	0,61	0,68	0,59	0,61
16	Information Technology	0,61	0,51	0,58	0,60	0,54	0,63	0,61	0,58
17	Television & Radio	0,58	0,50	0,52	0,53	0,50	0,56	0,58	0,54
18	Video	0,63	0,50	0,55	0,58	0,55	0,62	0,62	0,58
Interest in Creative Industries		0,61	0,55	0,56	0,57	0,54	0,58	0,59	0,57

Based on the above table can be obtained information that the sub-sectors of the most popular creative industry of the seven objects of research are the music and photography sector with a percentage of choosing is 66%, followed by 63% design and art, craft, and culinary with a percentage of 61 %. This shows the trend among students in filling the creative industry sub-sector leading to the music and photography sector. Another case with a sub-sector that is less desirable. The above tabulation results show that the fashion sector is less desirable by students of technology and engineering with a percentage of 48%. When viewed from the average of all respondents of the study obtained 57% interest in student creative industries. If referring to the result of student interest to creative industry, the biggest value is owned by SVHS 3 Yogyakarta which in fact is the state school with high enough value that is 61% while the lowest is PVHS Muhammadiyah 1 Bantul Regency which is 54%.

The search is more comprehensive by providing the three most popular choices of students. The results obtained as presented in Figure 2 below show that the most desirable sub-sector of students in school is photography. The dominance of this sector compared to other sectors gives researchers speculation that the majority of students are using a smartphone to capture the moment through photos and then published to social media. While the second sequence is filled by the music industry sector, and the next is the culinary sector followed by the design sector. The survey results of three alternatives to the selection of creative industry sub-sectors that are least desirable are the sector of publishing

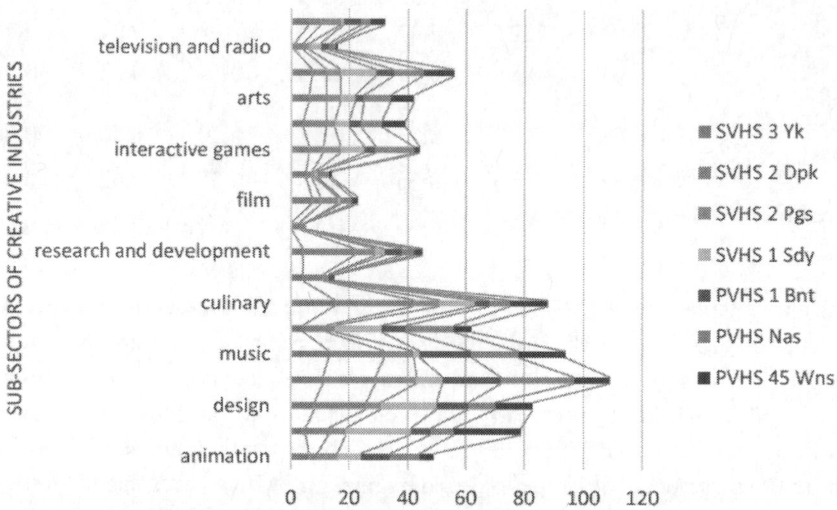

Figure 1. Selected sub-sectors based on priority scale

Interest searches and factors influencing the selection of creative industry sub-sector were done by socialization method. The great enthusiasm of respondents in participating in the socialization of creative industries in VHS shows that creative industry programs are highly anticipated for action to be followed up by schools in order not to open self-development that is not in tune with the potential of the majority of VHS students. In exploring why students choose the priority sub-sectors, researchers use matching instruments based on what they choose. The results of matching sub-sector interests with the reasons for choosing can be seen in the following table:

Table 2. Reasons for selecting creative industry priority sub-sectors

Reasons for choosing	Aggregate
Passion	54
Expectations	122
Hobby	243
Environment	54
Personal needs	47
Family wishes	9
Job demands	16
Supporting facilities	20
Follow the trend	21
Existence	34
Curiosity	209
Etc.	16

IOP Conf. Series: Materials Science and Engineering **535** (2019) 012023 doi:10.1088/1757-899X/535/1/012023

Based on these results, the hobby is a factor that dominates VHS students to select the creative industry sub-sector of their choice. Then the next highest score is the sense of student curiosity and student expectations of the sub-sectors of his choice. When viewed in the table above, the expectation factor of parents to direct their children to work in the creative industry is very small compared to other factors. It is, of course, a record to all of us that in this era of openness students have a broad perspective to look at their bright future based on what they want not because of the will of the parents.

4. Discussion and Conclusion

The creative industry should indeed be introduced early to the students about the scope of work to become a new labor market for VHS graduates [9, 10]. The importance of the creative industry for VHS students provides an opportunity for those who want to develop their talents and hobbies toward the creative industries sub-sector. The results of this study also provide findings that the reasons students choose the sub-sector of the creative industry because it is dominated by hobbies or passion that they do in filling their daily activities. This indicates that the sustainability of the achievement of the creative industry program into the activities in the school is not independent of what the students are fond of. Very difficult to apply if students can do the development of sub-sector insight if not included with the daily penchant.

This research is limited to reveal the potential of VHS in the field of technology and engineering expertise. Consideration of researcher uses this field because creative industry in Indonesia still very layman for them. In their daily activities in the more activities toward engineering that provides a reference for researchers to explore whether there is another potential outside techniques that can support local creative industries. From the results of the tabulation of the research indicates that there is considerable potential for VHS students to explore the world of creative industries.

The creative industry sub-sectors that most students choose are photography and music, although the data comprehensively shows the photography sector that students are passionate about. The results of the research interviews during the data collection as a confirmation of the results reinforce that the reason they choose photography as an option because almost 95% of students in the class have a smartphone. They use a smartphone camera to capture the moment to upload to social media or youtube channels. This shows that there has been a trend called narcissism where all smartphone users will use their handphones to take pictures as desired [11].

Another sub-sector of interest is the music industry. This sector is widely developed by VHS through extracurricular music, both art gamelan, chorus, and band. The tendency of students much interested in this sector because students can hone their intelligence and emotional to poured into the music [12]. Sectors other than music that VHS students are not interested in is a mode. Random interview results show the reasons for the lack of interest in the sector because the sector is more likely to be fashionable where VHS students are less attention to his appearance when on the move. The reasons are supported because the population of this study is dominated by the gender of men who in fact choose a sector that tends to masculine.

The results of research on the interests of students in each sub-sector need to be detailed by making three levels of priority that should be selected respondents. The result of the instrument of interest based on the priority of choice is obtained that the reason they liked the industrial sub-sector of choice due to the hobby.

This means hobby becomes the most important part of directing students to deepen its ability in sub-sectors of choice. But the problem for students is the school facilities that are not ready to meet the needs of the sub-sector based on the student's choice. This is certainly a barrier to the creation of qualified creative industry resources.

The second reason is the students' curiosity towards the sub-sectors of the creative industry. This curiosity that is able to encourage students to love the sub-sector of creative industries of his choice. While the reason for the smallest students in the creative industry sub-sector is the expectation of parents. This is very contradictory because today's children learn faster to develop their talents and interests without the influence of parents. Access and information disclosure factors give students much less fulfillment of what future parents expect in relation to the filling of creative industries sub-sectors.

5. References

[1] Pangestu, M. E. (2015). *EKONOMI KREATIF INDONESIA 2025*. Jakarta: Departemen Perdagangan RI.
[2] Maryunani, S. R., & Mirzanti, I. R. (2015). The Development of Entrepreneurship in Creative Industries with Reference to Bandung as a Creative City. *Procedia - Social and Behavioral Sciences*, *169*(August 2014), 387–394. https://doi.org/10.1016/j.sbspro.2015.01.324
[3] Bekraf. (2017). 10 Kota Kreatif di Indonesia (2).
[4] ICCI. (2017). KADIN Dorong Ekonomi Kreatif Yogyakarta
[5] Suparmin, S., Roniwijaya, P., Priyanto, S., & Setiadi, B. R. (2017). Eksplorasi sub-sub sektor industri kreatif di pusat-pusat keramaian kabupaten kulon progo 1234. *Prosiding Seminar Nasional Multi Disiplin Ilmu*, 712–720.
[6] Setiadi, B. R., & Suparmin. (2017a). Providing Space for Creative Industry Sub-Sectors to Meet the Labor Market for Vocational School Graduates, *102*(Ictvt), 309–313.
[7] Howkins, J. (2013). *The Creative Economy: How People Make Money from Ideas*. Penguin UK.
[8] DVHSD. (2017). 10/8/2017 Data Pokok SMK.
[9] Rodgers, J. (2015). Jobs for creatives outside the creative industries: A study of creatives working in the Australian manufacturing industry. *Creative Industries Journal*, *8*(1), 3–23. https://doi.org/10.1080/17510694.2015.1034572
[10] Setiadi, B. R., & Suparmin, S. (2017b). Providing space of. *Procedding ICTVT UNY.*
[11] Clarke, I. E., Karlov, L., & Neale, N. J. (2015). The many faces of narcissism: Narcissism factors and their predictive utility. *Personality and Individual Differences*, *81*, 90–95. https://doi.org/10.1016/j.paid.2014.11.021
[12] Berk, R. A. (2008). Use of Technology and Music to Improve Learning. *The Role of Technology in Tomorrow's Schools*, 32. Retrieved from

IOP Conf. Series: Materials Science and Engineering **535** (2019) 012024 doi:10.1088/1757-899X/535/1/012024

The effects of a teaching factory learning model on students' competency: an empirical study at the department of automotive engineering education of Universitas Negeri Yogyakarta

B Sulistyo[1], I Siswanto[2], Tawardjono[3], and N Widodo[4]

Yogyakarta State University, Yogyakarta, Indonesia

E-mail: bambang_sulistyo@uny.ac.id

Abstract. Vocational education is expected to serve the function of preparing competent and skilled graduates to meet the needs of industries. Therefore, this study aimed to determine the effects of implementing a teaching factory learning model integrated into the subject of Vehicle Body Painting on the students' competency. Using the research design of an experimental approach, this study involved two research groups that consisted of Automotive Engineering students following the subject of Vehicle Body Painting. The control and the experimental groups were treated respectively with a training based learning model and a teaching factory learning model. Indicators to be observed were the results of the competency test and the feasibility of the students' work. The data were collected through a competency test and documentation. The findings of this study suggested that the teaching factory learning model had significant effects on the students' competency (83.82) compared to the students treated with the training-based learning model (77.87).

1. Introduction

The Directorate of Vocational Secondary Education which is also known as Direktorat Pembinaan SMK/ PSMK has the vision, as stated in the roadmap of the Directorate General of PSMK 2010-2014, to establish Vocational High Schools that produce passionate, intelligent, competitive graduates with national identity and entrepreneurial spirit who are able to develop local excellence and to compete in the global market (Directorate of Vocational Education, 2009). To achieve this vision, the formulated mission are improving access, expansion and equity in high quality Vocational High Schools for all levels of society; improving the quality of Vocational High Schools through the application of disciplinary attitudes, noble character, and environmentally friendly, ICT-centered, and contextual based learning; empowering Vocational High Schools to produce entrepreneurial and skilled graduates by working collaboratively with relevant industries and business entities in the form of "teaching industry" (PSMK, 2009). In the roadmap of SMK 2010-2014, it is targeted that at the end of 2014, 70% of Vocational High Schools has a business learning unit in the form of teaching industry or teaching factory (PSMK, 2009).

Teaching factory is a learning activity where students directly perform production activities either in the form of goods or services within the school environment (Indrayana, 2007; Isnandar, 2008). The resulted goods or services are of very high quality thus they are worth selling and accepted by costumers (Triatmoko, 2009). This benefits the school by increasing the source of funding for the school to support educational activities (PSMK, 2008). Teaching factory presents the real world of industries within the school setting to produce work-ready graduates (Chryssolouris, Mavrikios, & Mourtzis, 2013; Sampurno & Siswanto, 2015). Furthermore, teaching factory is also very compatible with the philosophy of vocational education delivered by prosser (1950: 217) that vocational school will be effective if the learning process is conducted in an environment that is a replica of the actual work environment.

The main goal of vocational education is to produce graduates equipped with expertise and competence in accordance with industry needs (Agrawal, 2013). However, the teaching factory as a supportive program in achieving this goal, in reality, faces many difficulties in its implementation. Siswanto (2011) reported a study in 8 favorite Vocational High Schools in Yogyakarta that revealed that Vocational High Schools in Yogyakarta had difficulties in the implementation of teaching factory. These difficulties have mostly occurred in the field of technology and tourism. For example, Vocational High Schools for technology including SMKN 2 Yogyakarta, SMKN 2 Depok Sleman, and SMKN 2 Wonosari and Vocational High Schools for Tourism that consisted of SMKN 5 Yogyakarta and SMKN 4 Yogyakarta.

Based on the background of the problem above, a model or an example of teaching factory implementation in vocational schools was required. This model was implemented in the Department of Automotive Engineering Education, Universitas Negeri Yogyakarta in the subject of Vehicle Body Repair and Paint. This subject was selected since the competency taught in the subject meet the needs of the industries with an average suitability level of 86.75% (Tawardjono, Widodo, Siswanto, & Sulistyo, 2016). While the achievement of students' competency in this subject was categorized as very good with the number of students with scores above B were 37 students (84,09%) in 2013 and 70 students (88,61%) in 2014 respectively (Tawardjono, et al., 2016).

2. Method

This study was designed with a quasi-experimental approach since it is not possible to control all the variables that might influence the students' learning outcomes (Cohen, Manion, & Morrison, 2007). It was intentionally conducted by giving certain treatment to the esearch subject to analyze the impact or effect of the treatment (Fraenkel & Wallen, 2008). This was in accordance with the objective of this study to determine the effect of the teaching factory model on the students' competency in the subject of Vehicle Body Repair and Paint.

Population in this study included all students in the Department of Automotive Engineering Education in the entry year of 2014 that consisted of 80 students. The students were divided into 2 groups, the experimental group, and the control group. The number of students in the experimental and the control groups were 41 and 39 respectively.

This study was conducted in the Department of Automotive Engineering Education, Faculty of Engineering UNY. This study also involved practitioners from the industry. The research design is presented in Table 1.

Table 1. Research Design

Group	Pretest	Treatment	Posttest
Control	-	X1	T1
Experimental	-	X2	T1

whare: X is Training based learning model; Y is Teaching factory learning model.

The research variables consisted of free variable and dependent variable. The free variable consisted of teaching factory learning model and training-based learning model. While the dependent variable was the student's competency in the subject of Vehicle Body Repair and Paint.

Data collection techniques were a competency test and documentation. The competency test assessed the effects of the implementation of the learning models to the improvement of students' competency. Documentation was used to determine the learning models implementation process. The data were analyzed descriptively to describe the implementation of the learning process and the students' achievement. In addition, an inferential analysis was performed to test the research hypothesis with anava.

3. Result and Discussion

3.1 Result

3.1.1 Preparation.

In the preparation stage, a cooperation agreement with industry partners was established. The class for the experimental test of the teaching factory teaching model was prepared. The industry partners consisted of car body repair and paint workshops located in Yogyakarta area. The experiment was conducted on bachelor program's students in the Department of Automotive Engineering Education UNY. It was performed by dividing the class into 2 groups, the experiment group, and the control group. The experimental group followed the learning process according to the developed teaching factory model. In contrast, the control group followed the learning process in accordance with the existing learning method where the students did practice sessions using tools and materials without any demands to meet costumers' needs.

To ensure the validity of the experiment, the following steps were performed: (1) Both groups got the same subject (2) The students of both groups had completed the same number of credits so they were expected of having the same initial ability to follow the subject of Vehicle Body Repair and Paint (2) The evaluation system for both groups used the same instrument.

3.1.2 Implementation.

Students who did the vehicle painting practice with the teaching factory model were divided into 8 groups with 5 to 6 members each group. The learning model was implemented with a block system. Each group did the practice sessions in the industry in accordance with a predetermined schedule. It was carried out in the effective period of the semester so that the students who followed this program had to leave lectures and practice sessions in the campus.

Practice sessions in the industry included handling costumers' complaints, estimating the process and the cost of the repair, and do the process of vehicle body painting. Within 1 semester, each group carried out a practice session for 3 x 8 hours.

As for the control group, the students performed the practice session in the workshop of the department. The control group was also divided into 8 groups with 4 to 5 members. The students used practical materials in the form of one vehicle that had been prepared to be used as the practical tool. Each group of students got the responsibility to do the painting process on one part of the vehicle. The painting process was done for 3 x 8 hours.

3.1.3 Evaluation.

Evaluation is an important aspect of the learning process (Bowman, 2015). The purpose of the evaluation is to obtain information on overall learning outcomes in accordance with the design (Bergstrom, 2011). The evaluation of the implementation of the teaching factory teaching model in this painting practice was conducted through an 8-hour competency test. Students performed painting process to a panel of a vehicle. Based on the result of the competency test, the achievement data of students' competency showed that the mean score of ME experimental group was 83.816 higher than the mean score of the control group with MK: 77.868 (ME> MK). Thus it proposed that learning with teaching factory was better than the conventional learning method confirmed by the significant difference calculated using SPSS of t = 3.547 with 99% confidence level.

Supporting and inhibiting factors in implementing the teaching factory learning model in the subject of body repair and paint.

In its implementation, there were supportive factors including (1) Cooperation from the industry who provided the location for the Teaching factory activities and bought products produced by the students; (2) The ease of collecting the necessary data, especially about the achievement of students' competency in the subject of body repair and paint; and (3) Support and assistance from lecturers who taught the subject of body repair and paint. In general, the implementation of the study ran smoothly in accordance with the plan, but there was one obstacle. The obstacle was the students had to leave several subjects because of the implementation of the Teaching factory with a block system. The implementation of a block system was carried out by sending each group that consisted of 5 students to work on teaching factory products in the effective period of the semester. To overcome this problem, the step were to provide socialization for the lecturers in charge of those subjects and to strongly suggest the students make up the lessons that were missed due to absence at a later point.

3.2 Discussion

The implemented learning model was a learning model developed through a preliminary study, a focus group discussion, and also had been presented in an international seminar (Tawardjono, et al., 2016). This learning model helps students to achieve higher learning outcomes than students treated with regular learning models because the students gain real experience in the workforce according to expected learning targets (Jonassen, 2009, Lamancusa, 2008). This is also supported by previous studies conducted by Rentzos, Doukas, Mavrikios, Mourtzis, and Chryssolouris (2014) which reported that teaching factory provides benefits to students by giving better outcomes to the project provided by the teacher. In addition, the students in the experimental group also gained experience in working in groups, received input from instructors directly in the workshop, as well as used industry standard practical tools and materials that support the learning objectives so that the achievement was higher than the students in the control group (Inayat, ul Amin, Inayat, & Salim, 2013).

The teaching factory learning model can be conducted properly if there is support from the industry. This is also in accordance with the previous study by Hanushek, Schwerdt, Woessmann, and Zhang (2015) which stated that countries that combine vocational education with apprenticeship programs in industries are able to produce graduates who are more adaptable to the world of work. A successful apprenticeship program is resulted from supports of the industry to accommodate and provide opportunities for students to learn directly in the world of work (Siswanto, 2012). Therefore the purpose of vocational education to prepare graduates with a certain degree in accordance with the needs of professional careers in the future can be achieved (Lunev, et al., 2016).

4. Conclusion

This study suggests a teaching factory learning model in the field of vehicle body repair and painting developed in the form of painting practices integrated with activities in a workshop and directly related to customers. This model is able to improve the students' competency. The implementation of a teaching factory learning model will be more effective with a block system, therefore the school should arrange a learning schedule that enables the implementation of teaching factory with a block system without reducing the time allocation for other subjects.

5. References

[1] Agrawal, T. (2013). Vocational education and training programs (VET): An Asian perspective. *Asia-Pacific Journal of Cooperative Education, 14*(1), 15-26
[2] Bergstrom, P. (2011). Shifting the emphasis from teaching to learning: process-based assessment in nurse education. The international review of research in open and distance learning, 12 (5)
[3] Bowman, S.F. (2015). Evaluation in instructional design practice: a view from the stakeholders (Doctoral dissertations). Available from ProQuest Digital Dissertation and Theses database. (UMI No. 3712122)
[4] Chryssolouris, G., Mavrikios, D., & Mourtzis, D. (2013). Manufacturing systems: skills & competencies for the future. *Procedia CIRp, 7*, 17-24
[5] Cohen, L., Manion, L., & Morrison, K. (2007). The ethics of educational and social research.In Research Methods in Education. Sixth edition. London: Routledge, 51- 77.
[6] Direktorat PSMK. (Mei 2008). *Kewirausahaan dalam kurikulam SMK.* Makalah disajikan dalam Seminar Nasional Wirausaha Kuliner, di Jurusan Teknologi Industri, Fakultas Teknik , Universitas Negeri Malang
[7] Direktorat PSMK. (2009). Roadmap pengembangan SMK 2010-2014. Jakarta: Departemen Pendidikan Nasional
[8] Fraenkel, J. R. & Wallen, N. E. (2008). *Introduction to qualitative research: How to design and evaluate research in education* (7th ed). Boston, MA: McGraw-Hill International Edition
[9] Hanushek, E. A., Schwerdt, G., Woessmann, L., & Zhang, L. (2017). General education, vocational education, and labor-market outcomes over the lifecycle. *Journal of Human Resources, 52*(1), 48-87
[10] Inayat, I., ul Amin, R., Inayat, Z., & Salim, S. S. (2013). Effects of collaborative web based vocational education and training (VET) on learning outcomes. *Computers & education, 68*, 153-166
[11] Indrayana, D. (2007). *Kota vokasi ciptakan efek ganda.* Accessed on 16 August 2010 at http://dedi-indrayana.blogspot.com
[12] Isnandar. (Mei 2008). *Penyiapan entrepreneurship melalui pembelajaran teaching factory.* Makalah disajikan dalam Seminar Gelar Cipta Boga di Hotel Kartika Graha Malang.
[13] Jonassen, D.H. (2009). Learning and instructional technologies for the 21st century. USA: Springer
[14] Lamancusa, J.S. et al. (2006). The learning factory: industry-partnered active learning (online). *Journal of engineering education,* 97, 1
[15] Lunev, A. N., Safin, R. S., Korchagin, E. A., Sharafutdinov, D. K., Suchkova, T. V., Kurzaeva, L. V., Nikishina, S.R., & Kuznetsova, N. A. (2016). The mechanism of industrial educational clusters creation as managerial entities of vocational education. *International Review of Management and Marketing, 6*(2S)

IOP Conf. Series: Materials Science and Engineering **535** (2019) 012024 doi:10.1088/1757-899X/535/1/012024

[16] Prosser, C.A. & Ouigley, T.H. (1950). Vocational education in a democracy (revised edition). Chicago, USA. CA: American technical society.

[17] Rentzos, L., Doukas, M., Mavrikios, D., Mourtzis, D., & Chryssolouris, G. (2014). Integrating manufacturing education with industrial practice using teaching factory paradigm: A construction equipment application. *Procedia CiRP*, *17*, 189-194.

[18] Sampurno, Y. G. & Siswanto, I. (2015). Teaching factory di SMK Muhammadiyah 2 Borobudur Magelang. *Auto tech-pendidikan teknik otomotif, 6(1).*

[19] Siswanto, B.T. (2012). Pengaruh model penyelenggaraan work-based learning rolling terpadu terhadap aspek-aspek kualitas hasil belajar. Jurnal pendidikan dan teknologi kejuruan, 21 (1), 67-73.

[20] Siswanto, I. (2011). Pelaksanaan teaching factory di SMK RSBI Daerah Istimewa Yogyakarta. Thesis. Universitas Negeri Yogyakarta.

[21] Tawardjono, U., Widodo, N., Siswanto, I., & Sulistyo, B. (2016). Teaching factory learning process at vehicle's body repairing and painting workshop of faculty of engineering Yogyakarta State University. Prosiding International Conference on Educational Research and Innovation (ICERI).

[22] Triatmoko, SJ. (2009). The ATMI story, rainbow of excellence. Surakarta: Atmipress.

Home-made PLC laboratory training set for limited budget vocational education institution

Sunomo[1] and A Asmara[2]

[1,2]Department of Electrical Engineering Education, Yogyakarta State University, Yogyakarta, Indonesia

E-mail: sunomo@uny.ac.id

Abstract. The objective of this study is to build a low voltage three phase 24/41-volt programmable logic controller (PLC) lab. experiment training set for vocational education institution with limited budged. The training set is very cheap, durable and save from electric shock hazard due to no other and no higher voltage used for running it. The 12 I-O PLC unit using ATMEGA 16 processor, programmed by LD micro ladder diagram software on the laptop connected to the PLC module. The loads are: two units three phase motor modified from washing machine motor with specific designed, three phase electronic magnetic contactor, three modified motorcycle bulb lamp and one buzzer. The training set equipped with two version protection; electronic phase failure for the three phase source and electronic fuse destroyer in every unit module. The 24/41 volt three phase source will be blackout one second after one of the 8 ampere three phase fuse become off by overload condition or destroy due to a wrong connection of the individual module from phase to neutral (24 volt) become phase to phase(41volt). The evaluation result from the three internal department of PLC programmer expert and five vocational secondary school (SMK) teachers in Yogyakarta shows that the unit works well as it intended.

1. Introduction

Programmable Logic Controller (PLC) is a commonly used for lab experiment course in vocational education due to application of automation system in modern industry. PLCs are covered in many undergraduate automation and control-related courses, such as; Computer-Aided Manufacturing, Computer-Integrated Manufacturing, Industrial Control and Networking, Sensors and Instrumentation, Manufacturing Automation and Robotics [1]. For education, Rana proposed a Universal Simplest possible PLC using Personal Computer, The PLC unit has been produced by industries eg; ABB, Siemens, Omron, Mitsubishi, Hitachi, GE, Fanuc, Schneider, Zelio etc, and standardized for industries applications [2]. For educational purposes, PLC Training set can be ordered from Festo didactic, Allen-Bradley, Shandong Xinke China, SCE Siemens and others. But if the price is a problem due to limited budget, a vocational education institution can make a PLC training set module with electric load to be controlled, for example; very low voltage three phase motor, buzzer (alarm), lamp, led module traffic light arranged. With LDmicro, the PLC itself can be built by ordinary microcontroller and programmed using LDmicro free software prepared by Jonathan Westhues Portland, Oregon [3].

In this study, PLC training set build with 12 input-output ATMEGA 16 processor, and with electric loads to be controlled is; two very low voltages three phase motor 24/41 volts via electronic magnetic contactor, buzzer, and bulb lamp. The motor itself made by modification of secondhand washing machine capacitor motor, the bulb lamp constructed from double filament 12 volts' motorbike lamp, in series with 12,5 ohm 40 watt to reduce the light intensity and its current. The buzzer is DC 6-24volt specification, series connected with a diode and resistor 220 ohm to make it fed with 24-volt ac. With the whole system operated in the same voltage 24-volt ac for input and output terminal, the price will be low and ideal for institutions which have a limited budged. No different voltage sources are needed to perform the experiment on board like the module proposed by [4]. Optocoupler is used to bridge the different voltage inside the circuits of microcontroller,

2. Method

The method using in this study is R & D, by designing the physical form and testing its performance. All of the performance tested running in within four hours. The evaluation of its appropriateness done with open sistem questionnaire to three people PLC programmer in the department and five people from vocational high school (SMK) is; SMKN2 Yogyakarta, SMKN3 Yogyakarta, SMKN 1 Sedayu Bantul, SMKN1 Pundong, SMK YAPPI Wonosari.

2.1 The Whole Module

The physical form of the whole module shown in Figure 1a-1b. With laptop connected to the PLC module via USB to make a program via ladder diagram (Fig.1a) and running the program with laptop connection released from the PLC module (Fig.1b). The bottom side of the module (main module on Fig.1a) is 3 phase 24/41 volts source with three 220V/24 volt 5 ampere transformers connected in star to convert 220/380 volts source to 24/41 volt, protected with MCB in the primer and switched on and off by magnetic contactor. In the secondary circuit (24/41 volt), protection done by 5 ampere fuse on each phase and electronic three phase failure relay. If one of the 5 ampere fuse is off, it takes one second for the 24/41-volt source will be blackout. The physical modules of the PLC and the magnetic contactor seen from rear side shows in fig.2a and 2b, Fig 3a is the three phase motor modules seen from the rear side, laid on the upper side of the main 24/41 three phase source box, connected via sockets to the acrylic motor symbol terminal. Fig.3b is the circuit module of phase failure protector (left) and over voltage fuse destroyer (right). No screw used for every individual module, just insert and play on the main slot power modul with the motor laid on the upper side plywood box where the three power transformer 10 ampere 220volt/24 volt installed in the box (see Fig 1a and Fig 3a).

IOP Conf. Series: Materials Science and Engineering **535** (2019) 012025 doi:10.1088/1757-899X/535/1/012025

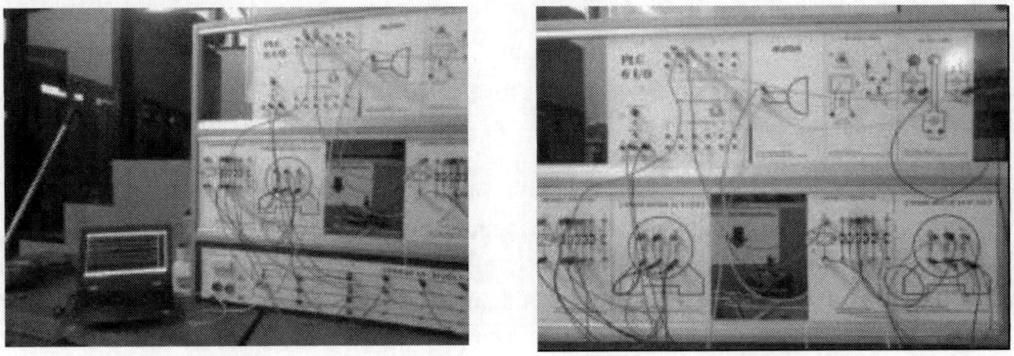

(a) (b)

Figure 1. The PLC training set module ; **(a)** programming via laptop **(b)** running condition

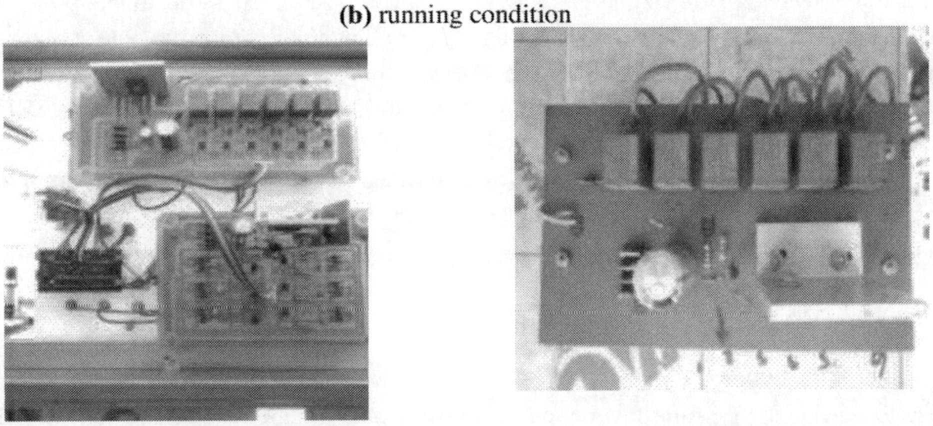

(a) (b)

Figure 2. The physical module seen from the rear side ; **(a)** The PLC **(b)** The Contactor magnetic

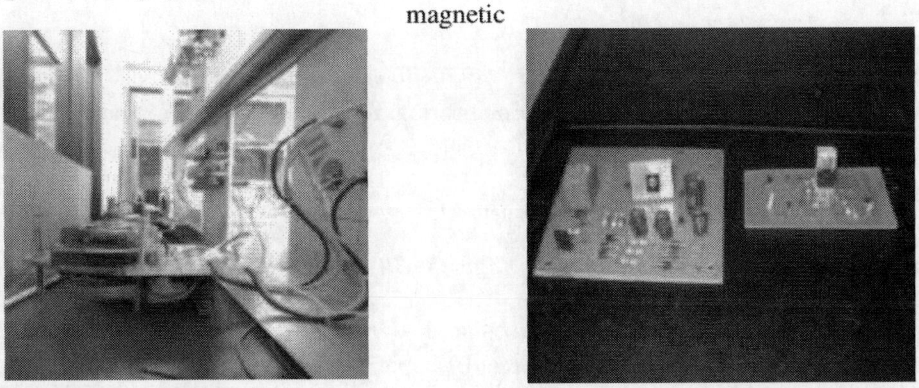

(a) (b)

Figure 3. The physical module seen from the rear side ; **(a)** 24/41 three phase motors, **(b)** Phase failure and over voltage electronic protector units

2.2 Current measurement for the loads

The measurement result of current draws by the loads are: 1,03 amperes for motor in star connection and 0,97 – 1,05 amperes for the bulb lamps, 0,30 amperes for electronic magnetic contactor, and 0,0346 amperes for buzzer. Due to the easy circuit

2.3 Outlines of loads and protection design

For good skills of vocational education teacher or lecturer in basic electronics and electrical machines, home-made loads described in this paper is easy to implemented. To convert the motor from one phase to four pole in three phase mode, taking the data about number of slots, diameter of wire and turn per coil of the main winding coil in 220volt, then converted to turn per coil in linear comparison for 24 volt, and use the ordinary rewinding formula available for the phase mode. In this study, single layer winding was applied. For the electronic magnetic contactors, use the ordinary 12 volt dc relays with its contact poles arranged as ordinary three phase contactors. To energized from 24 volt ac, use the ac to dc simple power supply that acts as magnetic coils in three phase contactor. To avoid wrong connection from phase to neutral 24Vac) become phase to phase (41Vac) done by anyone who operate the unit, use the TRIAC fuse destroyer with peak back-to back zanier diode sensor setting at little more of 24 volt ac peak voltage. This will be save the filter capacitor and series pass transistor breakdown in the dc power supply in the electronic magnetic contactor. Electronic phase failure relay 24/41 volt connected in series with stop push button of the main magnetic contactor in the main module that convert the 220/380volts ac main source to 24/41 volt via 3x 1 phase 5A or 10 ampere transformer.

3. Conclusion

The evaluation results from the eight evaluators shows that the whole PLC lab experiment training set work properly as it intended, recommendation or suggestion only in the standardization of the electrical schematic, whether it will follow totally to the ANSI, IEC or PUIL (Indonesia) standard.

4. References

[1] Sheng-Jen (Tony) Hsieh 2003 *Animations and Intelligent Tutoring Systems for Programmable Logic Controller Education* (Department of Engineering Technology and Industrial Distribution, Texas A&M University) Int. J. Engng Ed. Vol. 19, No. 2, pp. 282±296

[2] B.K.Rana *Universal Simplest possible PLC using Personal Computer Electrical Engg* (India: VSSUT)

[3] Westhues, Jonathan 2017 *LDmicro: Ladder Logic for PIC and AVR* (http://cq.cx/ladder.pl access March 14, 2017).

[4] Erdal Yilmaz , Sevan Katrancioglu 2003 *Designing Programmable Logic Controller (PLC) Experiment Set with Internal Experiment Blocks* (Great Britain: Elsevier Ltd) ; Procedia - Social and WCETR 2011Behavioral Sciences 28 (2011) 494 – 498 1877-0428

IOP Conf. Series: Materials Science and Engineering **535** (2019) 012026 doi:10.1088/1757-899X/535/1/012026

Machining industry's contribution level in vocational high school revitalization

A E Wibowo[1] and S Munadi[2]

[1]Graduate School, Yogyakarta State University, Yogyakarta, Indonesia
[2]Department of Mechanical Engineering Education, Faculty of Engineering, Yogyakarta State University, Yogyakarta, Indonesia

E-mail: sudji.munadi@uny.ac.id

Abstract. Revitalization of Vocational High Schools (VHS) is an effort to increase the role of all stakeholders in developing education. This study aimed to measure the level of machining industry contribution in VHS revitalization, especially in Machining Skills Competence in Yogyakarta Special Region (DIY). This study used descriptive research with quantitative approaches. The population was 20 machining industry in Yogyakarta Special Region. Samples were determined by saturated sampling techniques. The number of samples selected were 14 machining industries. Questionnaires were used to measure the level of the machining industry contribution in implementation of VHS revitalization. Instrument validation used logical and empirical validity tests. Quantitative data analysis techniques used descriptive analysis. The results of the study showed that the level of machining industry contribution in VHS's revitalization, especially in Machining Skills Competence in DIY as a whole was in the "very low" category with a score of 42.67%.

1. Introduction

In Indonesia, skilled intermediate labour is produced through Vocational High Schools (VHS). VHS are educational institutions that prepare graduates to work in certain fields. VHS graduates are expected to be a productive workforce. Ideal VHS graduates are workers who are ready to use, in the sense that they can directly work in the business / industry or entrepreneurship. This is stated in Law number 20 of 2003 concerning the National Education System Article 18 and an explanation of Article 15 which regulates Vocational Secondary Education. However, in reality there are still many SMK that have not been able to produce the ready to use labour according to the industry needs so that many SMK graduates still do not get a job.

Head of the Central Statistics Agency (BPS), revealed that from the Open Unemployment of 5.50 percent in August 2017, the highest unemployment was a VHS graduate of 11.41 percent. According to the Social Statistics Deputy of BPS, the number of unemployed vocational school graduates due to the expertise of SMK graduates is not necessarily in accordance with the needs of the company. Agreeing with this reason, the Head of the National Development Planning Agency assesses something is wrong with the vocational education curriculum in Indonesia so that the skills taught in schools are not in accordance with the needs of today's business world, where many new jobs are emerging that require certain skills. Therefore, the Indonesian government focuses on improving the quality of human resources (HR) through vocational education as a national priority in 2018.

One of the efforts made by the government to improve the quality of vocational schools is the Vocational Revitalization program. The legal basis for this revitalization is Presidential Instruction No.

IOP Conf. Series: Materials Science and Engineering **535** (2019) 012026 doi:10.1088/1757-899X/535/1/012026

9 of 2016 concerning Revitalization of Vocational Schools. Strategic issues which are the priority of vocational revitalization, namely updating curriculum alignment with industrial needs; learning innovation; fulfilment and improvement of the professionalism of teachers and education personnel; partnership / school collaboration with the business world / industrial world (DUDI) and universities; standardization of key facilities and infrastructure; and institutional arrangement / management. The strategic issue means that it has become imperative, that the key to the success of vocational revitalization is the collaboration between the school and the industry. This is in accordance with the statement delivered by the Minister of Education when launching an industrial vocational education program that every vocational school must partner with industry.

One of the efforts made by the government to improve the quality of vocational schools is the Vocational Revitalization program. The legal basis for this revitalization is Presidential Instruction No. 9 of 2016 concerning Revitalization of Vocational Schools. Strategic issues which are the priority of vocational revitalization, namely updating curriculum alignment with industrial needs; learning innovation; fulfilment and improvement of the professionalism of teachers and education personnel; partnership / school collaboration with the business world / industrial world (DUDI) and universities; standardization of key facilities and infrastructure; and institutional arrangement / management. The strategic issue means that it has become imperative, that the key to the success of vocational revitalization is the collaboration between the school and the industry. This is in accordance with the statement delivered by the Minister of Education when launching an industrial vocational education program that every vocational school must partner with industry.

Education Revitalization is an effort to increase the role of all stakeholders in developing education. According to Hadam, et al., revitalization of Education is a more thorough effort, more persistent and more responsible to realize the goal of national education [1]. Revitalization in the context of vocational education is intended to maximize the role of the Government, Regional Government, and the World of Work / Industrial World to really care for Vocational Schools. Vocational revitalization is carried out through several steps, but if observed there are 6 steps that involve the industry's contribution as follows. 1) Cooperation "link and match"; 2) industrial work practices; 3) establishment of industrial class; 4) the implementation of an internship teacher (on the job training); 5) Industrial Based Curriculum; 6) Factory Teaching.

Based on the description above, it is necessary to study DUDI readiness in implementing the VHS revitalization, especially in machining engineering expertise. The readiness can be seen from the level of industry's contribution to VHS development as a foundation in determining strategic steps to strengthen the relationship between vocational schools and machining industries.

2. Method

This study used descriptive research with quantitative approaches. The population was 20 machining industry in Yogyakarta Special Region. The decision of selecting the samples is through saturation sampling method because of the small population. The number of samples selected were 14 machining industries. Questionnaires were used to measure the level of the machining industry contribution. Instrument validation used logical and empirical validity tests. Quantitative data analysis techniques used descriptive analysis.

3. Result and Discussion

The contribution level of the machining industry in the overall SMK revitalization effort is in the very low category. This can be seen from 14 respondents, the contribution rate of 9 respondents is in the very low category and the contribution rate of 5 respondents is in the low category. In addition, when viewed from the accumulation of scores and the average accumulation of scores, the level of contribution can be expressed in percentages by comparing the mean value of the research results by

47.79 with an ideal score of 112, so that the contribution rate is 42.67%. The interpretation of the percentage level of contribution of the machining industry in DIY in the vocational revitalization efforts, as a whole, falls into the "very low" category.

These findings further emphasize that the contribution of the machining industry in DIY in order to participate in implementing the steps to revitalize vocational schools needs to be improved. This finding is in accordance with the results of research conducted by Indriaturrahmi which say that the role of related industry in implementing regional government policies to the SMK development is not yet optimal [2]. The Head of Department of Education, Youth and Sports DIY Department of Education also added that the contribution of the business world and industry (DU-DI) in the development of vocational secondary education is still low, so the DIKPORA DIY Department conducts various activities in order to facilitate the improvement of access to education to the industrial world.

These findings deserve to be a basic evaluation by the government in this case the relevant institution (Department of Industry and Department of Education), vocational school, and machining industries. Government policy will greatly influence the role of industry in participating in developing vocational schools. Schools must also improve to improve good relations through partnerships and mutually beneficial cooperation with industry, so that the industry is more comfortable to participate in developing vocational schools. The results of contribution level analysis of the machining industry in vocational revitalization efforts in each aspect can be seen in table 1.

Table 1. The Contributions Level of Machining Industry in VHS Revitalization Efforts

No	Aspect of Contributions	Contributions level (%)	Categories
1	Cooperation aspect for building "link & match"	55,60	Low
2	Apprenticeship program	72,15	High
3	Teacher internship program	31,45	Very low
4	Industrial class program	25	Very low
5	Curriculum development	36,61	Very low
6	Teaching factory program	32,14	Very low
	Over all	42,67	Very low

3.1 The contribution level of machining industry in cooperation aspect for building "link & match"

The contribution level of the machining industry to cooperation aspect is expressed in percentages which is done by comparing the average value of the research results of 15.57 with an ideal score of 28 so that the contribution level of 55.60% is obtained. Interpretation of the percentage level of contribution of the machining industry in DIY in the effort to revitalize vocational schools in aspects of cooperation falls into the "low" category. This is in line with the results of research by Bambang and Budi which states that the results of the partnership between SMK and industry are still not optimal, so that the placement of graduates in work is not in accordance with the competencies of students [3]. Azizah, Murniati, and Khairuddin, in their research recommended that schools and stakeholders be more communicative in establishing cooperative relationships to improve graduates' competencies [4].

3.2 The contribution level of machining industry in apprenticeship program

The contribution level of machining industry to aspects of the apprenticeship program is expressed in percentages which is done by comparing the average value of the research results of 14.43 with an ideal maximum score of 20 so that the contribution rate of 72.15% is obtained. The interpretation of the percentage level of contribution of the machining industry in DIY in the efforts to revitalize vocational schools in aspects of industrial work practice programs is in the "high" category. This is an important capital for the development of vocational education. The high contribution of industry to the apprenticeship program is inseparable from the reason that the industry can get a positive impact directly through this program. The results of interviews with 14 industry respondents showed that 6 industries said they were greatly helped by the existence of students. This is in line with the opinion of Hee who said that industrial work practices programs can benefit the industry because they can obtain the fresh ideas and energy that students bring into the work environment [5].

The findings above also show the weaknesses that have occurred during the implementation of industrial work practices, namely cooperation to coordinate and formulate industrial work practice training programs between the vocational school and industry is still very low. This is in accordance with the study conducted by Rasyid which states that industries that are school partners have not been able to participate in planning student learning activities in shaping student professionalism [6]. These findings further emphasize that students do work practices without planning so that sometimes students do not get learning in accordance with the competencies that must be mastered.

Collaboration to coordinate and compile instruments for evaluating work practices between the SMK and industry is still relatively low. Evaluation instruments are only arranged unilaterally by the school. Pragmatically, the school can compile apprenticeship assessment instruments that will be used widely in each industry where apprenticeship works, but it is important to realize that the competencies taught in each machining industry vary. Even in some industry sources, they said that they composed their own assessment instruments according to their ability. Furthermore, they said that they found it difficult to conduct an assessment because they did not have the ability in the field.

It is appropriate if both parties sit together to coordinate the things that will be assessed considering that industrial work practices are a collaborative program between schools and industry. Sudjana said that the success of revealing students' results and learning processes as they are (objectivity of assessment results) depends very much on the quality of the assessment tools in addition to the way they are implemented [7]. Therefore, it is expected that both parties can coordinate to formulate and use the instrument and include this activity as one of the points in the memorandum of understanding on technical internship.

3.3 The contribution level of machining industry in teacher internship program

The level of contribution of the machining industry to the aspect of the teacher apprenticeship program is expressed in percentages which is done by comparing the average value of the research results of 6.29 with an ideal maximum score of 20 so that the contribution level of 31.45% is obtained. The interpretation of the percentage level of contribution of the machining industry in DIY in the efforts to revitalize vocational schools in the aspect of the teacher apprenticeship program is in the "very low" category. These findings indicate that collaboration between Vocational Schools in particular Machining Engineering Skills Competencies with the machining industry needs to be improved. This is important because the teacher internship program can improve and update the teacher's understanding of the latest technology used in the industry and needs to be understood by students. Billet believes that the type of material delivered by a teacher to his students depends on his understanding and beliefs [8]. This implies that if a teacher has never taken an industrial apprenticeship program, the lesson he prepares for the class will not be in line with industry development. Therefore the pattern of cooperation and the development model of industrial-based productive vocational teacher competencies needs to be developed.

3.4 The contribution level of machining industry in industrial class program

The contribution level of the machining industry to the aspect of the industrial class program is expressed in terms of percentage which is done by comparing the average value of the research results by 6 with a maximum score of 24 so that the contribution rate of 25% is obtained. The interpretation of the percentage level of contribution of the machining industry in DIY in the efforts to revitalize vocational schools in aspects of industrial class programs is in the category of "very low".

These findings indicate that there is no machining industry in DIY in collaboration with vocational schools to carry out industrial class programs. These findings also show that industrial class programs have not been implemented in vocational schools, especially in Machining Engineering Skills Competencies in DIY. This is certainly contrary to the government's considerable support. The government, through the Directorate of Vocational Development, has provided funding for the implementation of industrial class programs.

According to Yoto, based on the results of his study, steps that can be taken to prepare for the industrial education class are as follows: (1) building harmonious cooperation with industrial partners where apprenticeship; (2) planning an industrial education class (industrial education) model with industry which is outlined in a cooperation agreement by both parties; (3) compile a joint curriculum according to industry needs; (4) determine the needs of teachers / instructors who teach in schools and in industry; (5) determine the facilities and infrastructure of practices, textbooks and learning resources that must be prepared at school by both parties; (6) determine the learning schedule at school and industry (on the job training); (7) determine the implementation of national exams and competency tests [9].

3.5 The contribution level of machining industry in curriculum development

The contribution level of the machining industry in the aspect of curriculum development is expressed in percentages which is done by comparing the mean value of the research results of 2.93 with an ideal maximum score of 8 so that the contribution rate of 36.61% is obtained. The interpretation of the percentage level of contribution of the machining industry in DIY in the effort to revitalize vocational schools in the aspect of curriculum development is in the category of "very low".

These findings are in accordance with the findings of Mohtadi et.al which states that 52% of the opinions of workers and 64% of academic opinions state that the industry does not contribute maximally in reducing the competency gap between the world of education and the industrial world [10]. Industry contribution can be increased if the communication and cooperation relations between the two are improved. López and Pérez argue that the relationship between vocational education and the world of work holds an essential key to which vocational education must be able to provide education that is able to meet the needs of the workforce [11]. The findings of Mohtadi et.al, also state that, that academics and industry believe that gaps can be bridged through collaboration between stakeholders [10].

3.6 The contribution level of machining industry in teaching factory program

The contribution level of machining industry in the aspect of the teaching factory program is expressed in percentages which is done by comparing the average value of the research results of 2.57 with a maximum score of 8 so that the contribution rate of 32.14% is obtained. The interpretation of the percentage level of contribution of the machining industry in DIY in the SMK revitalization efforts in aspects of the factory teaching program is in the "very low" category. These findings indicate the low level of cooperation between SMKs and related industries. Siswanto released the results of the study that most of the RSBI Vocational Schools in DIY have not cooperated with industry related to the implementation of the teaching factory. This implies that over a decade, vocational cooperation with industry has become an unresolved classical problem [12].

4. Conclusions

The contribution level of the machining industry in vocational revitalization efforts, as a whole is included in the "very low" category with a score of 42.67%. The results of this study can be used as a reflection for policy makers, industry and vocational schools in improving vocational school revitalization efforts. In addition, more in-depth research is needed to explore the inhibiting and supporting factors for machining industry in contributing to the vocational revitalization program. The inhibiting and supporting factors can then be used as a basis for improvement in the future.

5. References

[1] Hadam S, Nastiti R., & Ayu N.A. (2016). Strategi Implementasi Revitalisasi SMK. Jakarta: DPSMK.

[2] Indriaturrahmi. (2016). Peran Dunia Usaha dan Dunia Industri dalam Penyelenggaraan SMK Berbasis Kearifan Lokal di kota Mataram. *Jurnal Pendidikan Vokasi,* 6, 162-172.

[3] Ixtiarto, B. dan Sutrisno, B. (2016). Kemitraan Sekolah Menengah Kejuruan dengan Dunia Usaha dan Dunia Industri (Kajian aspek Pengelolaan Pada SMK Muhammadiyah 2 Wuryantoro Kabupaten Wonogiri) [Electronic version]. *Jurnal Pendidikan Ilmu Sosial,* Vol 26, No.1, Juni 2016, 57-69.

[4] Azizah, Murniati A.R., dan Khairuddin. (2015). Strategi Kerjasama Sekolah dengan Dunia Usaha dan Dunia Industri (DUDI) dalam Meningkatkan Kompetensi Lulusan pada SMK Negeri 3 Banda Aceh. *Jurnal Administrasi Pendidikan,* 3, 148 – 158.

[5] Hee, K.Y. (2012). Industrial Training Benefits Students and Companies. Taken on 9 Agustus 2018 dari https://www.swinburne.edu.my/campus-beyond/industrial-training-benefits-students-and-companies.php

[6] Rasyid, M. 2008. Dukungan Industri terhadap Keberhasilan Pendidikan Sistem Ganda di Sumatera Barat. *Forum PendidikanUNP,* No. 01 Tahun XXIII, 53-67.

[7] Sudjana, N. (2009). *Penilaian Hasil Proses Belajar Mengajar.* Bandung: PT. Remaja Rosdakarya.

[8] Billet, S. (2011). *Vocational Education: Purposes, Traditions and Prospects.* New York: Springer.

[9] Yoto. (2014). Model "Diklastri" Sebagai Alternatif Meningkatkan Mutu Lulusan SMK. *Jurnal Pendidikan Sains.* Vol.2, No.3, 125-131.

[10] Mohtadi C., McAree O., & Schlosser J. (2014). *Bridging the Skills Gap in STEM Industries.* Paper presented in 42nd Annual Conference, di Birmingham,UK.

[11] López, B.G., & Pérez, C.P. (2014). Transversal Competences for Employment and Profile of Excellent University Students. *Procedia-Social and Behavioural Sciences.* 139, 305-313.

[12] Siswanto, I. (2008). *Faktor Pendukung dan Penghambat Pelaksanaan Teaching Factory di SMK RSBI Daerah Istimewa Yogyakarta.* Tesis Master, unpublished, Universitas Negeri Yogyakarta, Yogyakarta.

IOP Conf. Series: Materials Science and Engineering **535** (2019) 012027 doi:10.1088/1757-899X/535/1/012027

Developing project based learning module of CNC milling mechanical technique on mechanical engineering department vocational high schools in Surakarta

T U Nugroho [1] and T Sukardi [2]

[1, 2] Technology and Vocational Education, Postgraduate of Yogyakarta State University, Yogyakarta, Indonesia

E-mail: triujann@gmail.com

Abstract. This development research that use ADDIE model with stages: (1) Analysis, (2) Design, (3) Development, (4) Implementation, and (5) Evaluation. The object of research was students of XII class on Vocational High Schools in Surakarta. Internal trials included validation material expert and media expert, while external trials included teachers and students responses to use module. The trials data was collected through questionnaire, observation, documentation, and interview which were analyzed qualitatively and quantitatively. The research result: (1) The CNC learning module developed is suitable for CNC learning, students can learn independently and thoroughly, includes description and CNC learning goals, adaptived to development of science and technology, students easy to read and understand; (2) CNC learning module developed were declared very feasible by material expert feasibility percentage of 83.2% and media expert feasibility percentage of 84.04%, the teachers response to the module is 95.6% and the students response is 85.8% which module is very suitable to be used; (3) increasing of students learning outcome after using the module is 17.09% and student learning completeness reaches 100%. The conclusion of research results is module very feasible and effective to be used for CNC learning at Vocational High Schools in Surakarta.

1. Introduction

The current of globalization is changing lifestyle of society from traditional agrarian to modern industry. Indonesia participation in Trends in International Mathematics and Science Study (TIMSS) and Program for International Student Assessment (PISA) since 1999 to show that the achievement of Indonesian childrens are not so good. This is because the number of test materials in TIMSS and PISA is not found in the education curriculum in Indonesia [1]. The loss of Indonesia in the education sector has encouraged the government to organize and develop an international standard education unit at all level of education. The path of education held in Indonesia consists of formal, nonformal, and informal education. The formal education level consists of basic education, secondary education, and higher education, while the types of education composed general education, professional and vocational education, religious education, and special education [2].

Vocational High School is held to improve intelligence, knowledge, personality, noble character, and skills of students to live independently and follow further education in accordance with their vocational program [3]. In order to work optimally, they must have strong stamina, master the basic of science and technology, have a high work ethic, be able to communicate according to work demand, and have the ability to develop themselves. Vocational High Schools emphasize that graduates have competencies that are in accordance with their field of expertise, for example in the Mechanical Engineering Department, one of the competencies that must be mastered by students is Computer Numerical Control (CNC). This competency is very important, because many industries now use CNC on based production machines to achieve higher productivity. Competency standard in CNC Machining Technique is to require students to master three competency standards, that is setting NC/CNC machine, programming NC/CNC machine, and operating NC/CNC machine [4].

CNC Learning for Vocational High Schools has not been able to be implemented in accordance with the applicable curriculum objectives, due to the lack of details of competency achievement indicators in core competencies and basic competencies, and the limited facilities and infrastructure owned by Vocational High Schools. As a comparison of ratio the number of CNC machines to the number of practicums at Vocational High Schools in Surakarta, where one CNC machine is used for 10-20 students. Based on data from the Student Skills Competition in Surakarta on 2017 for CNC competition, there are only three participants, that is State Vocational High School 2, St. Mikael Vocational High School, and Warga Vocational High School [5], whereas in Surakarta there are around 12 Vocational High Schools that have Mechanical Engineering Department. The low participation of Vocational High Schools participating in Student Skills Competition is due to the lack of ability of Vocational High Schools to prepare CNC machining competency, both in terms of teaching materials, learning component, and infrastructure facilities.

The CNC learning process at Vocational High Schools in Surakarta which includes material, learning strategies, infrastructure, and learning schedules is not sufficient to achieve student competency, due to the absence of standard CNC learning processes for students. The Learning Implementation Plan made by the teacher still refers to the cognitive or theoretical domain, while in the psychomotor domain or practice there is still less. In learning CNC, a teacher who is competent and able to teach the material thoroughly to students is needed to master all CNC competencies taught [6]. However, there are still many teachers who are not yet competent, such as not being able to operate CNC machine in their schools and the low ability of teachers to use computer software as a simulation media and CNC programming. In addition, it is also due to the lack of teaching materials and learning media that students can use for independent learning, such as module and CNC simulator. Therefore, it is necessary to do research on the Development Learning Module of CNC Milling Machining Technique with the help of CAD/CAM software that achieves through Project Based Learning in the Mechanical Engineering Department at Vocational High Schools in Surakarta.

Table 1. Data Analysis of CNC Learning Observation at Vocational High Schools in Surakarta

Nu.	Vocational Name	Curriculum	CNC Module	Number of CNC	Simulation Machines
1	SMK Negeri 2	2013 Curriculum	√	12	Mastercam
2	SMK Negeri 5	2013 Curriculum	√	11	Mastercam
3	SMK Bhineka Karya	KTSP	-	2	Mastercam
4	SMK Kristen 2	KTSP	√	3	Mastercam
5	SMK Muhammadiyah 1	2013 Curriculum	-	1	Mastercam
6	SMK Murni 1	KTSP	-	1	-
7	SMK Pancasila	2013 Curriculum	-	2	Mastercam
8	SMK PGRI 1	2013 Curriculum	-	2	Mastercam
9	SMK St. Mikael	KTSP	√	8	Mastercam
10	SMK Tunas Pembangunan 2	KTSP	-	1	-
11	SMK Tunas Pembangunan 3	KTSP	-	-	-
12	SMK Warga	2013 Curriculum	√	13	Mastercam

Learning activities require teaching materials as the delivery of material to students. One of the to used teaching materials as reference is module [7], which is expected to be able to streamline communication and interaction between teachers and students in CNC learning. The research results of Wijanarka (2012) show that the development of learning module effective to shaping students competencies in accordance with the demands of the curriculum and all CNC competency standards can be achieved by students after applying and using module [8]. Based on the results of preliminary study at Vocational High Schools in Surakarta, there are some teachers who do not use the module when delivering CNC learning. Although there are, the material discussed in module is not accordance with the type of CNC machines in the Vocational High Schools.

Learning media that can be used to support CNC learning are computer software, such as Computer Aided Manufacturing (CAM), which has graphic facilities for simulating machining before being implemented on real machines to prevent errors during the CNC machining process. The research results of Abdulrasool (2010) indicate that CNC learning process with the help of computer software and CAD/CAM modules can improve effectiveness and learning outcomes when compared to manual/conventional learning [9]. With computer assisted learning, students are expected to be actively involved in learning, so teacher centered learning changes to student centered learning [10].

In addition to infrastructure, teaching materials, and media learning, the learning strategies used by teachers in teaching can also support the achievement of learning goals [11]. The research results of Yalcin (2009) show a statistical difference between the experimental group students taught by the PjBL method and the control group [12]. Some teachers of CNC learning at Vocational High Schools in Surakarta still teach using conventional learning models and teacher centered. The selection of learning strategies must be based on consideration that students as learning subjects have experience, knowledge, and desire to learn, and have confidence that they are capable and able to learn. Learning strategies are needed to integrate students theoretical and practical abilities, so that they can achieve the overall CNC Machining Technique competencies.

2. Method

The research used is Educational Research and Development (R & D), the research method used to produce or develop an educational product which is then tested for its feasibility and effectiveness [13]. This type of development research is designed using ADDIE model development research which consists of five steps, that are: (1) Analyze; (2) Design; (3) Development; (4) Implementation; and (5) Evaluation. The research on the development of the ADDIE model was chosen because it has a complete development stage, as well as instructions and development steps arranged in a sequential and systematic manner [14].

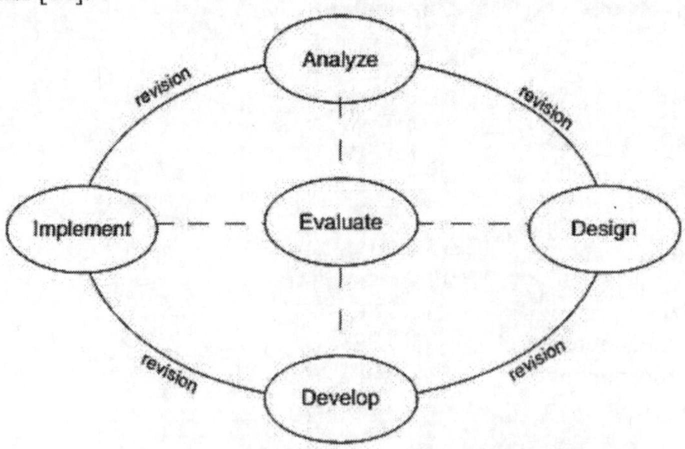

Figure 1. ADDIE Development Model [15]

Application of modules is carried out through internal trials by expert and external trials which can be said as field test. Data for external trials evaluation can be done through pretest-posttest, and if possible interview [16]. Data analysis obtained from the results development of the ADDIE model was carried out descriptively. The questionnaire was analyzed to provide an overview of the quality of the CNC learning module that was developed and the improvement of the module based on the input and suggestions given. Descriptive analysis of qualitative data was carried out on data from preliminary studies, while descriptive analysis of quantitative data was carried out through statistical calculations to analyze product trial results data.

3. Results

3.1 Results of Expert Appraisal

Module must be assessed first by experts consisting of material expert and media expert to determine whether the module is ready to be tested in the field.

3.1.1 Material Expert Assessment from the Education. The results of material expert assessment from education obtained an average score of 3.98 with good module criteria, and percentage of module feasibility showed 79.62% with module qualifications feasible to be used.

Table 2. Results of Recapitulation Material Expert Questionnaire from Education

No.	Aspect of Evaluation	Amount of Item Score	Average Score	Percentage Score	Feasibility Level
1	Feasibility of Module Contents	13	3,9	78,46 %	Feasible
2	Grammar	11	4	80 %	Feasible
3	Presentation of Module Contents	9	4	80 %	Feasible
4	Module Benefits	7	4	80 %	Feasible
	Average Assessment Score		3,98	79,62 %	Feasible

3.1.2 Material Expert Assessment from Industry. The results of material expert assessment from industrial world obtained an average score of 4.34 with excellent module criteria, and percentage of module feasibility showed 86.95% with module qualifications very feasible to be used.

Table 3. Results of Recapitulation Material Expert Questionnaires from Industry

No.	Aspect of Evaluation	Amount of Item Score	Average Score	Percentage Score	Feasibility Level
1	Feasibility of Module Contents	13	4,3	86 %	Very Feasible
2	Grammar	11	4,27	85,4 %	Very Feasible
3	Presentation of Module Contents	9	4,11	82,2 %	Very Feasible
4	Module Benefits	7	4,71	94,2 %	Very Feasible
	Average Assessment Score		4,34	86,95%	Very Feasible

3.1.3 Learning Media Expert Assessment. The results of learning media expert assessment obtained an average score of 4.2 with excellent module criteria, and percentage of module feasibility showed 84.04% with module qualifications very feasible to be used.

Table 4. Recapitulation of Learning Media Experts Questionnaire

No.	Aspect of Evaluation	Amount of Item Score	Average Score	Percentage Score	Feasibility Level
1	Module Size	4	4,25	85%	Very Feasible
2	Module Cover Design	11	4,27	85,55%	Very Feasible
3	Module Contents Design	25	4,08	81,6%	Very Feasible
Average Assessment Score			4,2	84,04%	Very Feasible

3.2 Field Trial Results
Field trials were conducted twice, that are small group trials and large group trials using the experimental method.

3.2.1 Small Group Trials. Small group trials were carried out in one of Vocational High Schools in Surakarta which had CNC machining facilities that were in accordance with the modules developed, that is Warga Vocational High School. Based on the results of filling in student response questionnaires obtained an average score of 4.06 with good module criteria and percentage of module feasibility show 81.2% with module qualifications feasible to be used.

Table 5. Results of Student Response on Small Group Trials

No.	Aspect of Evaluation	Amount of Item Score	Average Score	Percentage Score	Feasibility Level
1	Presentation of Material	15	4,04	80,8 %	Feasible
2	Grammar	7	4,27	85,4 %	Very Feasible
3	Display/Graphics	6	3,92	78,4 %	Feasible
4	Module Benefits	12	4,03	80,6 %	Feasible
Average Assessment Score			4,06	81,2 %	Feasible

3.2.2 Large Group Trials. Large group trials were conducted on two Vocational High Schools which were considered able to represent other Vocational High Schools in Surakarta, that are Surakarta State Vocational High School 2 and Warga Vocational High School. The application of modules in the two Vocational High Schools uses an experimental approach with aim to determine the feasibility and effectiveness of module by comparing how much learning increases between experimental class or class using modules with control classes or classes that do not use modules (10). Data from the results of large group trials show that average value of experimental class students in both the cognitive and psychomotor domains is higher than the value of the control class students.

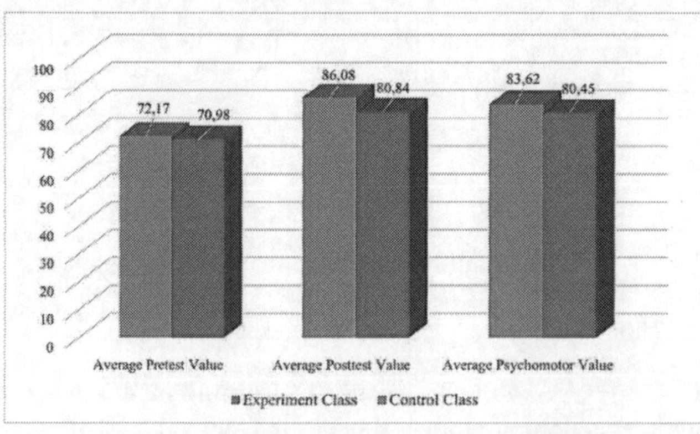

Figure 2. Comparative Diagram of Value Students of Experimental Class with Control Class at Surakarta State Vocational High School 2

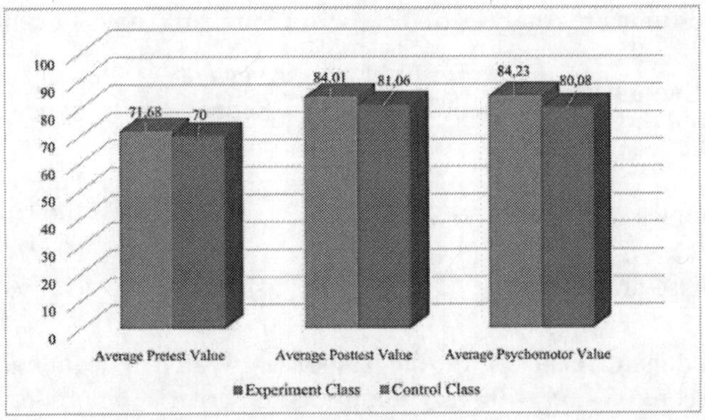

Figure 3. Comparative Diagram of Value Students of Experimental Class with Control Class at Warga Vocational High School in Surakarta

The results of questionnaire analysis at Surakarta State Vocational High School 2 students showed an average score of 4.26 with very good module criteria and percentage of module feasibility showed 85.35% with module qualifications very feasible to be used in CNC learning.

Table 6. The Questionnaire Analysis of Students in Surakarta State Vocational High School 2

No.	Aspect of Evaluation	Amount of Item Score	Average Score	Percentage Score	Feasibility Level
1	Presentation of Material	15	4,25	85%	Very Feasible
2	Grammar	7	4,29	85,8%	Very Feasible
3	Display/Graphics	6	4,30	86%	Very Feasible
4	Module Benefits	12	4,23	84,6%	Very Feasible
	Average Assessment Score		4,26	85,35%	Very Feasible

The results of questionnaire analysis at Warga Vocational High School students showed an average score of 4.32 with very good module criteria, and percentage of module feasibility showed 86.4% with module qualifications very feasible to be used in CNC learning.

Table 7. The Questionnaire Analysis of Students in Warga Vocational High School

No.	Aspect of Evaluation	Amount of Item Score	Average Score	Percentage Score	Feasibility Level
1	Presentation of Material	15	4,39	87,8%	Very Feasible
2	Grammar	7	4,23	84,6%	Very Feasible
3	Display/Graphics	6	4,36	87,2%	Very Feasible
4	Module Benefits	12	4,30	86%	Very Feasible
	Average Assessment Score		4,32	86,4%	Very Feasible

The results of questionnaire analysis of the responses Surakarta State Vocational High School 2 teachers obtained an average score of 4.89 with very good module criteria, and module feasibility percentage was 97.9% with module qualifications very feasible to be used in CNC learning.

Table 8. The Questionnaire Analysis of Teacher in Surakarta State Vocational High School 2

No.	Aspect of Evaluation	Amount of Item Score	Average Score	Percentage Score	Feasibility Level
1	Presentation of Material	13	4,85	97 %	Very Feasible
2	Grammar	11	4,73	94,6 %	Very Feasible
3	Display/Graphics	9	5	100 %	Very Feasible
4	Module Benefits	7	5	100 %	Very Feasible
	Average Assessment Score		4,89	97,9 %	Very Feasible

The results of questionnaire analysis of the responses Warga Vocational High School teachers obtained an average score of 4.68 with very good module criteria, and module feasibility percentage was 93.7% with module qualifications very feasible to be used in CNC learning.

Table 9. The Questionnaire Analysis of Teacher in Warga Vocational High School

No.	Aspect of Evaluation	Amount of Item Score	Average Score	Percentage Score	Feasibility Level
1	Presentation of Material	13	4,46	89,23%	Very Feasible
2	Grammar	11	4,64	92,73%	Very Feasible
3	Display/Graphics	9	4,78	95,55%	Very Feasible
4	Module Benefits	7	4,86	97,14%	Very Feasible
	Average Assessment Score		4,68	93,7%	Very Feasible

4. Discussion

4.1 Analyze

In analysis phase a preliminary study was conducted on Vocational High Schools in Surakarta that have a Mechanical Engineering Department conducted through literature studies, observations, and interviews which then the results are translated into several aspects, including document analysis, learning analysis, analysis of observation results, analysis of books and internet, and analysis of interview results. At this stage a general analysis of the curriculum and aspects of learning is applied to Vocational High Schools in Surakarta. The curriculum analysis carried out cover core competencies, basic competencies, competency achievement indicators, and learning materials to determine which competencies require learning modules, for example in CNC Milling competency.

In analysis phase a preliminary study was also conducted on the CNC learning process to find out the learning needs as material to develop a learning module for CNC Milling Machining Techniques. Data needed in the form of completeness of infrastructure, teacher teaching process, modules, learning media, and learning strategies used. From the results of analysis, it was found that a problem must be followed up, such as the need for a CNC learning module and used as a material consideration for designing the design Project Based Learning Module of CNC Milling Machine Techniques.

4.2 Design

The design phase is development stage which is Project Based Learning Module of CNC Milling Machining Techniques using a scientific approach in accordance with 2013 Curriculum.

4.2.1 Module Framework. The learning module of CNC Milling Machining Techniques is arranged in sequence starting from the title page, cover, preface, table of contents, module position map, glossary, introduction, learning CNC, learning activities, closing, bibliography, and attachments. In the introductory section, the background for making the module is explained, the prerequisites students must have before using the module, instructions for using the module for both students and teachers, learning tools that support learning, the final goals to be achieved from the use of modules, core competencies and basic competencies, and checklists students initial abilities.

4.2.2 Systematics of Module Material Presentations. The systematic of module material presentation is based on core competencies and basic competencies in CNC Machining Techniques which have been translated into learning indicators. The material presentation in the module is divided into five learning activities, which in each learning activity contains learning objectives, material descriptions, and formative tests.

4.2.3 Module Assessment Instruments. The module assessment instrument consists of grids and several aspects of assessment. Module assessment is carried out by several parties, including assessment of material expert from the world of education and industry, learning media expert, CNC learning teachers, and students as respondents. Assessment of material expert from the world of education and industry includes aspects of the feasibility of module content, grammar, presentation of module content, and project based assessment. Learning media expert assessments cover aspects of module size, module cover design, and module content design. In addition, there is also an assessment from teachers which includes the feasibility aspects of the module content, grammar, presentation of the module contents, and the benefits of using the module. While assessment of students includes aspects of material presentation, grammar, appearance/graphics, and benefits of using module.

4.3 Development

Development is stage of producing product development which is carried out through internal validation and external validation. Internal validation is carried out by expert in their fields, while external validation is done by field trials consisting of small group trials and large group trials. At this stage of development several action modules were developed, including:

4.3.1 Draft Module Writing, consists of covers, introductory words, table of contents, module position map, glossary, introduction, learning activities (material, practice questions, job sheets, and evaluations), closing, bibliography, and attachments.

4.3.2 Module Editing, carried out by paying attention to module specifications, including in the form of printed media which are arranged in clear and easy to understand languages, and are based on the curriculum used in Vocational High Schools, that is 2013 Curriculum.

4.3.3 Module Validation, carried out in two stages namely internal validation and external validation. carried out in two stages that are internal validation and external validation. Whereas external validation is carried out by students and teachers after learning with modules. All of these validations were carried out to measure the feasibility level of module as learning teaching material for CNC Machining Techniques.

4.4 Implementation

The implementation for Project Based learning modules of CNC Milling Machining Techniques was carried out on XII class students of Mechanical Engineering Department at Vocational High Schools in Surakarta, namely in Surakarta State Vocational High School 2 and Warga Vocational High School. Both of these Vocational High Schools were chosen because they were considered to represent CNC learning at Vocational High Schools in Surakarta. In implementing this module learning students are trained to learn individually and in groups starting with discussions about material, looking for other learning resources, and working together in working on tasks and projects discussed in module. The implementation phase aims to determine students understanding in using CNC learning modules, as well as to determine level of readability of module and student responses regarding the modules used.

From the implementation of the module in small groups, some suggestions and inputs must be revised to make module better. After revision, module still needs to be implemented in a larger group, namely in the broader group of students or classes in several Vocational High Schools in Surakarta. In the implementation phase of a large group, a project based learning strategy is used to find out whether the revised module is already suitable for use in CNC learning at Vocational High Schools in Surakarta.

4.5 Evaluation

Evaluation of the results of product implementation shows that module is suitable for use in CNC learning, but only requires some repairs or revisions. Improvement include revisions to content, presentation/appearance, and others. Improvements made are efforts to improve module based on the results of assessment of expert, as well as the results of assessment and interviews with teachers and students. Improvements can be in the form of adding module supporters and not substance improvements. The conclusion that can be drawn at this stage is that CNC learning module developed is suitable for use in CNC learning.

Evaluation of development and use of modules is done through a scientific approach based on the results of evaluations from validators or experts, teachers, student responses, and the results of module implementation in small groups and large groups at Vocational High Schools in Surakarta. After the researcher has revised the evaluation phase, final product produced is a CNC learning module for Vocational High Schools students in Surakarta, whose validity has been tested and said to be appropriate as a teaching material, and known the effectiveness use of module.

5. Conclusion

Characteristics of Project Based Learning module of CNC Milling Machining Techniques developed according to CNC learning at Vocational High Schools in Surakarta, that are students can learn independently and thoroughly, including descriptions and CNC learning goals, have adaptive power to development of science and technology, and easy to read and understand students. The assessment of CNC Learning module based on material expert obtained an average score of 4.16 with good criteria and percentage of module feasibility showed 83.2% with module qualifications very feasible to be tested in CNC learning, while the assessment of CNC learning modules from media expert obtained an average score of 4.2 with very good criteria and percentage of module feasibility showed 84.04% with module qualifications very feasible to be tested in CNC learning. Assessment of modules by CNC learning teachers obtained an average score of 4.78 with very good criteria and percentage of module feasibility of 95.6% which indicates that module qualifications are very suitable for use in CNC learning, while response of students in use of modules obtained an average score of 4.29 with very good criteria and percentage of module feasibility 85.8% which indicates module qualifications are very feasible to be used as learning materials for CNC learning at Vocational High Schools in Surakarta. The research and development results show that Project Based Learning module of CNC Milling Machining Techniques is very helpful for students in improving their learning outcomes as evidenced increase student grades by 17.09%, and helping students completely CNC Machining Techniques as evidenced by 100% of students gaining value ≥ minimal completeness criteria.

6. References

[1] Kemendikbud. *(2013). Permendikbud RI Nomor 70, Tahun 2013, tentang Kerangka Dasar dan Struktur Kurikulum Sekolah Menengah Kejuruan.*

[2] Republik Indonesia. *(2003). Undang-Undang RI Nomor 20 Tahun 2003 tentang Sistem Pendidikan Nasional.*

[3] Kemendiknas. *(2006). Permendiknas RI Nomor 23, Tahun 2006, tentang Standar Kompetensi Lulusan untuk Satuan Pendidikan Dasar dan Menengah.*

[4] Kemendiknas. *(2009). Permendiknas RI Nomor 28, Tahun 2008, tentang Standar Proses Pendidikan Dasar dan Menengah.*

[5] Ditpsmk. (2017). *LKS CNC Milling Kota Surakarta.* Diakses pada tanggal 13 September 2015, dari: http://event.ditpsmk.net/lks2017/cnc-milling

[6] Arends, R.L. (2004). *Emploring teaching an introduction to education.* New York & London: Routledge Falmer.

[7] Komalasari, (2010). *Media pengembangan pembelajaran.* Jakarta: PT. Gramedia.

[8] Wijanarka, B.S. (2012). *Pengembangan modul dan pembelajaran kompetensi kejuruan teknik pemesinan CNC SMK.* Disertasi, Universitas Negeri Yogyakarta, Yogyakarta.

[9] Abdulrasool. (2010). *The Effect of Using Computer Technology Tools to Enhance the Teaching-Learning Process in CAD/CAM-CNC Module in Mechanical Engineering.* Journal of Information and Education Technology. Association for the Advancement of Computing in Education (AACE), ISBN 1-880094-79-7.

[10] Sanjaya, W. (2009). *Strategi pembelajaran berorientasi standar proses pendidikan.* Jakarta: Kencana.

[11] Joice, B. & Weil, M. (2004) *Models of teaching (7th ed.).* Boston: Pearson education.

[12] Yalcin, S.A., Turgut, U. & Buyukkasap, E. (2009). *The Effect of Project Based Learning on Science Undergraduates Learning of Electricity, Attitude Physics and Scientific Process Skills. Ataturk University, International Online Journal of Educational Sciences. No.1, pp. 81-105.*

[13] Sugiyono. (2013). *Metode Penelitian Pendidikan: Pendekatan Kuantitatif, Kualitatif, dan R & D.* Bandung: Alfabeta.

[14] Pribadi, B.A. (2009). *Pengembangan sistem pembelajaran.* Bandung: Pelita Ilmu.

[15] Branch, R.M. (2009). *Intructional design: The ADDIE approach.* University of Georgia.

[16] Dick, W. & Carey, J.O. (2005). *The systematic design of instruction (6th ed.).* Boston: Pearson Education Inc.

IOP Conf. Series: Materials Science and Engineering **535** (2019) 012028 doi:10.1088/1757-899X/535/1/012028

A comparative study of the occupational health and safety implementation in the industries and vocational high schools in Yogyakarta

B Styawan[1], T Sukardi[2], D Rahdiyanta[3], B S Wijanarka[4] and Y Ngadiyono[5]

[1,2,3,4,5]Graduate Program, Yogyakarta State University, Yogyakarta, Indonesia

E-mail: benny.styawan2016@student.uny.ac.id

Abstract. This study aims to describe the extent of the Occupational Health and Safety (OHS) implementation in the manufacturing industries compared with that in the Mechanical Engineering study program of the vocational high schools (VHSs) in Yogyakarta Special Region. It uses the qualitative descriptive statistical analysis with data collection techniques covering field studies with direct observation and interviews. The research participants, the employees of the manufacturing industries and the VHSs' Mechanical Engineering teachers in Yogyakarta Special Region, were determined purposively. The results show that the application of OHS in the industrial workplaces is somehow better than that in the VHS's Mechanical Engineering study program in the researched region. In general, the problems of the OHS application in VHSs are perhaps triggered by the lack of health and safety equipment, facilities, and infrastructure in the mechanical engineering workshops and the poor OHS management system (OHSMS). The recommendations made are that VHSs improve the OHSMS, conducting OHS education and training programs for the VHS teachers and staffs, and using rewards and sanctions for all of the school community members.

1. Introduction

Occupational health and safety (OHS) refers to the health aspect in the workplace and work environment. Admittedly, work safety can improve the efficiency of work performance and productivity. However, Indonesia still has a low level of work safety. Based on the data from the International Labor Organization (ILO) there were 99,000 cases of workplace accidents of which the total victims, about 70% have fatal consequences, death or permanent disabilities (ILO, 2012) [1]. Meanwhile, the Healthcare and Social Security Agency (*BPJS*) notes that the total number of the workplace accidents reached 123,000 cases in 2017 with the total claim value of Rp 971 billion, an increase of 20% from that in 2016 that accounted over Rp 792 billion (finance.detik.com, 2018). Thus, the OHS implementation in the industries aims to assure the health and safety of the workers and other people in the workplaces, and guarantee the production sources that they can be used safely and efficiently (Suma'mur, 2006) [2]. Finally, By minimizing of harm possibility, employees must act safety and safely (Adiratna et al, 2003) [3].

The companies that are aware of the importance of the employee's productivity will always pay attention to factors contributing to such productivity, one of which is done by implementing the OHS program. Despite the fact that the OHS implementation is the responsibility of all parties, the industrial communities and VHSs should involve and play an active role in the implementation of OHS culture to improve the quality of human resources and their productivity. Additionally Nielsen (2013) [4] in his research suggests that the health and safety organization (HSO) can improve the company's safety culture by focusing on safety-related interactions. This is in line with Rijuna Dewi (2006) [5] who states that OHS has a positive and significant effect on the employee performance. Based on Law No. 1 of 1970 on Occupational Health and Safety [6] and the Regulation of the Republic of Indonesian Government No. 50 of 2012 on the Implementation of the Occupational Health and Safety Management System (OHSMS) [7], OHS should applies in not only industrial workplaces and job agency but also schools and colleges. Thus, OHS needs to be applied in the practicums held by VHSs as well.

Concerning this, VHSs as the education and training providers function to generate skilled workers by offering a variety of study programs tailored to the needs of the world of work, one of which is the Mechanical Engineering. Often, the flagship programs inVHSs are the productive ones, which aim to equip students with the competencies based on the Indonesian National Working Competency Standards (*SKKNI*). Besides, to establish good work attitudes, strategies to improve the quality of OHS implementation through the learning process of OHS in schools should be carried out. According to Triadmidi (2010) [8] VHS graduates from the mechanical engineering study program will generally work as beginner workers or supervisors which deal directly with relatively dangerous tools, machines, and materials. Therefore, viewed from the existing problems, the objective of this paper is to describe the OHS implementation in industries and VHSs and to measure to what extent the OHS applies in either the manufactories or VHSs.

2. Method

This is a qualitative descriptive research study with the participants determined purposively.The manufacturing industries and the VHSs' mechanical engineering study programs in Yogyakarta Special Region were its setting. The data are in the forms of the results of field studies carried out through direct observation, in-depth interviews, and documentation. The data analysis was carried out by combining the primary and secondary data, categorizing them by a substantive theme prepared, rearranging them for the data interpretation, and drawing the conclusions. The validity test used the data triangulation so that the data presented are valid ones.

3. Results and Discussion

3.1. The OHS Implementation in the Industries

Occupational health and safety (OHS) resembles a system employed by an institution or company in regulating, supervising, and managing people working in the industrial workplaces. By monitoring and managing the working employees, it is expected that OHS can reduce the risks of occupational accidents.

In general, the health and safety programs in the industrial places cover several main aspects regarding the health and safety rules, rule awareness, protective clothing, floor cleanliness, enough light in the production areas, and work environment. Other aspects also include firefighting equipment, first aid and medical services, sufficient number of toilets, bathroom cleanliness, the programs, local directions, clean drinking water, health and safety committees, emergency procedures, and health and safety promotions. Of these aspects, however, this study focuses on the four aspects the OHS implementation in the manufacturing industries in Yogyakarta Special Region, namely, the aspects of work safety, occupational health, occupational health and safety management system (OHSMS), and competence.

In terms of work safety, the findings of the OHS application in the manufacturing industries located in Yogyakarta Special Region reveal that (a) the building and the arrangement of tools and machines apparently suit the applying standards. (b) The use of work equipment and the machine operation has also been somewhat in accordance with the standard operational procedure (SOP) applies in the company. Moreover, (c) the personal protective equipment (PPE) such as safety helmets, safety shoes, earplugs, work gloves, chemical respirators, dust masks, chemical googles, PVC gloves, welding gloves, welding masks, safety glasses, and the protective clothing seems to be complete. Besides, they also find that (d) the equipment and first aid kits apparently meet the standards stated in the Regulation of the Republic of Indonesia Minister of Manpower and Transmigration No. PER.15/MEN/VIII/2008 concerning the occupational first aid (OFA) [9]. Additionally, (e) the hydrants are available in every workroom, and (f) the environmental sanitation in the workplace is properly managed, proven by the cleanliness of the surrounding environment and the water quality, as well as the presence of industrial waste tanks. In overall, the OHS implementation in the manufacturing industries has been effective, but there are some problems, especially in terms of safety, including the management and utilization of waste products and the uninstalled alarms and emergency shower for fire detection in every room.

When viewed in terms of the occupational health, the OHS implementation in the industries covers the followings. (a) The work environment has met the Good cleanliness criteria, (b) noises are at the unwavering level, (c) the air conditioners or fans seem to be available in every workroom, and (d) enough light is assured in every workroom and hall as regulated. Besides, (e) clean culture in the workplaces has been carried out well such as the cleanliness of the stairs, floor, hallway, material storage, hand washing, and toilets, and the availability of landfills. Moreover, (f) the employees have been registered as the members of *BPJS Ketenagakerjaan* (social security programs for employers and individuals in Indonesia), and (g) there is support from the company leaders and good communication amongst workers. Based on these results, the OHS implementation in terms of the occupational health has run as planned. However, the findings find that such implementation has not run optimally as there are still many obstacles in the management of hazardous and toxic waste that has not been carried out well and the labeling for such waste is also not well coordinated.

Regarding the OHSMS, the manufacturing industries in the Yogyakarta Special Region have applied it effectively as governed by the Regulation of the Indonesian Minister of Public Works No. 05/MEN/1996 Chapter III Article 3 concerning the basic principles of OHSMS [10] that consist of five aspects carried out continuously. These aspects are (a) commitment/policy, (b) OHS planning, (c) implementation, (d) measurement/evaluation, and (e) review and improvement. Apart from these five aspects, some new findings are found. Some of them refer to the implementation of Law No.1 of 1970 Section VIII Article 12 concerning the reward and punishment that has not been carried out optimally due to either the lack of supervision from the institutions in charge of OHS regulation or the availability of OHS experts. These two problems result in another problem in the formulation and the implementation of OHS programs in the field.

Regarding the OHS implementation in terms of the workers' competence, in general it is found that all workers know the OHS programs, implement the programs, and participate in evaluating the programs. All of these are carried out in Good category. Based on the observation in the field, the workers possess the knowledge of the OHS programs, including those about (a) the first aid kits, (b) the proper use of PPE, (c) safety signs, (d) SOPs on fire extinguishing, (e) the socialization of the OHS programs, and (f) the audit of OHS programs. Therefore, to strive for the quality improvement of the OHS programs in the manufacturing industries, the companies should always upgrade their workers' competence by providing some training and simulation on the OHS implementation.

3.2. The OHS Implementation in VHSs

The OHS implementation in VHSs constitutes the management system that influences the level of the teacher and student performance. However, apparently the OHS in VHSs has not been implemented and coordinated optimally among the school management, teachers, and students. Thus, the quality of teacher performance in implementing OHS appears to be a benchmark for the OHS implementation in such schools.

In terms of work safety, based on the results of the observation in the mechanical engineering workshops, the OHS implementation in VHSs includes the followings. Firstly, the VHSs' workshop/laboratory building are in accordance with the Regulation of the Indonesian Minister of National Education No. 40 of 2008 concerning the facilities and infrastructure at VHSs [11]. Some of the machine maintenance is carried out periodically by cleaning the machine monthly, while its lubrication is carried out every 6 months, and the overhaul is carried out every 1 year. In addition, the OHS implementation in the mechanical workshops is not yet in accordance with the standards applied by the manufacturing industries. Moreover, there are many problems revealed from the observation of the OHS implementation in VHSs. They deal with the condition of the machines used for the practicums 50% of which are no longer appropriate to use. Besides, some of existing machines are damaged in its coolant, automatic component, and engine lighting. Such damage to the machines is due to either the aging factor orpoor maintenance.

In terms of the occupational health, the findings on the OHS application in VHSs' mechanical workshops uncover the presence of the School Health Clinic (SHC), lighting, ventilation, as well as the air circulation systems that function properly. However, there are still many problems found there. They are (a) the cleanliness of the workshops, hand washing, and toilets, (b) the workshops' wall stain, c) the provision of industrial waste disposal, and d) the presence of fans or blowers in the workshops.

Concerning the OHSMS in VHSs, the observation results show that this management system has not been implemented in the VHSs' workshops. The socialization of OHS programs has not been carried out effectively to teachers, staffs, and students. When referring to the basic aspects of the OHSMS, some staffs and teachers already know the OHS programs, but they do not implement them due to the ineffective communication among the teachers, staffs, students, and school management in formulating the OHS programs that should be implemented. Thus, it affects the implementation of the industry-based OHS culture in VHSs.

Despite the fact that VHS teachers have enough knowledge on the OHS, what hinder its implementation are the availability of OHS supporting facilities and infrastructure, and the ineffective communication among the school management, teachers, students, and staffs. These problems cause the teachers and staffs to ignore the OHS principles during the practicum process in the workshop, and later this condition is becoming a dilemma experienced by most vocational teachers.

3.3. A Comparison of the OHS Implementation in the Industries and in VHSs

There are some difference son how OHS applies in the industries and in VHSs, especially in terms of the facilities and infrastructure. The Regulation of the Indonesian Minister of National Education No. 40 of 2008 states that health and safety in the workshop scan be ascertained if sufficient PPE is provided to meet a certain minimum ratio of the total students. However, it is found that the available PPE in the VHSs' workshops is limited, and even the students sometimes should prepare their own PPE. When compared with that in the industry, the supply of PPE is quite enough or even more PPE is available for the number of employees working in one division, for example in the production section.

According to Article 1 concerning OFA for the workplace accidents, the first aid kits and the safety equipment condition are essential for the workers and other people who are at work who may experience pain or injury since the danger posed has been of high risks. Meanwhile, in VHSs the first aid kits are in poor condition and far from adequate for the students. Therefore, the OHS implementation in VHSs needs to be improved especially in the provision of OHS supporting facilities and infrastructure in the workshops.

In the Regulation of the Indonesian Minister of Public Works No. 09/PER/M/2008 [12], it is stated that the OHSMS is the part of the overall management system that includes the organizational structure, planning, responsibilities, implementation of procedures, processes and resources needed for development, implementation, achievement, study maintenance of OHS policies in order to control work-related risks to create a safe, efficient, and productive workplace. It is found that in the industrial workplaces such management system has been implemented in accordance with the applied regulation, but not in the VHSs. Based on data from the Indonesian Ministry of Education and Culture, the OHSMS was only applied to 60 State Vocational High Schools (*SMKN*) throughout Jakarta, Bogor, Depok, Tangerang, Bekasi (*JABODETABEK*) the contained of (Republika, 2016) [13]. Therefore, VHSs should prepare or develop an industry-based learning curriculum that enables them to apply OHS culture effectively and in the future, they can carry out the management system of the industry-based OHS by themselves.

In addition to the problem of OHS supporting facilities and the management system that distinguishes the application of OHS in industries and that in VHSs, the absence of OHS experts in VHS sand education and training for VHS teachers and staffs is another problem. In terms of the OHS knowledge, VHS teachers perform higher than do the industrial employees because the teachers' education is better than that of the employees in the industries. However, in terms of the OHS implementation, these employees has better implemented the OHS than have the teachers. Therefore, to balance this condition, VHSs should have work standards regarding the OHS implementation that are guided by that of the industries.

3.4. The Alternative OHS Programs for VHSs to Apply

Based on the description of the results and data analysis that has been provided in the previous sections, the important measures to carry out are as follows.

3.4.1. The OHS education and training. The OHS education and training for VHS teachers and staffs are in tended to improve their understanding on the importance of OHS programs. By applying the OHS programs in the workshop, it is expected that the risks of occupational accidents related to the machines used can be minimized and the productivity of teachers, staffs, and students may increase. Thus, the recommendation made is that VHSs should hold OHS education and training programs for teachers and staffs or establish cooperation with companies engaged in machining and manufacturing sectors to assist them in instilling the OHS culture.

3.4.2. Socialization on the OHS programs to the teachers, staffs, and students. Based on the findings, the socialization carried out by the school management to teachers, staffs, and students has not been properly carried out. These parties do not know and understand OHS programs, so that in practice, students often ignore the OHS. For example, students sometimes do not wear safety glasses and gloves when involved in the lathe practicum. Recommendation made is that the school management should participate in the formulation and management of the teaching and learning activities, especially in the productive courses carried out in the school with the aim that the school management can participate in the OHS implementation, supervision, and evaluation in the workshops.

3.4.3. The availability and completeness of personal protective equipment (PPE). The availability and completeness of PPE are the main issues in implementing the industry-based OHS. In this context, it is found that the PPE in VHSs is less complete, that is, some of the PPE is less feasible but it is still used for the practicum. Recommendations made are those concerning the PPE renewal, procurement, and maintenance

3.4.4. Health services at school. Health services in schools are commonly associated with those in the form of the SHC sand first aid kits. These health facilities and infrastructure are possessed by all schools, but for the VHSs in context, such SHC and first aid kits are considered less feasible or no longer appropriate to be used when accidents during the practicums happen. Besides, limited medicines and health tools become the next inhibiting factors when students are to use the first aid kits. Recommendations made are that the VHSs should provide the SHC and physicians, as well as a health insurance for teachers, staffs, and students so that they do not need to worry if accidents occur during the practicum in that the pain or injury can be directly handled by the phisicians.

3.4.5. Rewards and sanctions. This study found that rewards and sanctions have not run optimally in VHSs. In practice, rewards have not been given for teachers and staffs, but for students they are given as an additional value that is embedded to their productive learning achievement. Whereas for the sanctions, employees and teachers are not given any sanctions, but it is found that students are given warnings and fines when they cause some damage to the tools or components used for the practicum. It is recommended that rewards or sanctions should apply to not only students but also all school community members who are involved in the practicums done in the workshop. Thus, there will be no gap between teachers, staffs, and students in applying the work culture based on the industry-based OHS.

4. Conclusions

Based on the results of the data analysis, it can be concluded that the OHS implementation in the manufacturing industry is better than does that in the VHSs. The OHS implementation in VHSs still has many shortcomings, one of which is in the provision of facilities and infrastructure at school. These are caused by several problems, including (a) the rewards and sanctions that have not been implemented and (b) the provision of OHS facilities that is less optimal in terms of the procurement of PPE and first aid kids, and the cleanliness of the workshops. Moreover, (c) the OHSMS audit has not been scheduled by the school management, and (d) there is no OHS education and training for VHS teachers. The recommendations for the OHS implementation in VHSs concern with the procurement of tools, facilities, and infrastructure in accordance with Minister of National Education Regulation No. 40 of 2008, support for the OHSMS within the school scope, OHS training program for teachers and staffs, and the rewards and sanctions given for all school members.

IOP Conf. Series: Materials Science and Engineering **535** (2019) 012028 doi:10.1088/1757-899X/535/1/012028

5. References

[1] International Labour Organization 2012. *Global Estimate of Forced Labour: Result and Methodology, Special Action Programe to Combat Forced Labour*. (Geneva: ILO).

[2] Saut David 2018 February 6 *Angka Kecelakaan Kerja Meningkat ke 123 Ribu Kasus di 2017.* https://finance.detik.com/moneter/d-3853101/angka-kecelakaan-kerja-ri-meningkat-ke-123-ribu-kasus-di-2017 on 28 July 2018.

[3] Suma'mur, P K 2006 *Keselamatan Kerja dan Pencegahan Kecelakaan* (Jakarta: PT. Gunung Agung)

[4] Adiratna Y, Tsani M H FX, and Djumialdi 2003. *Pelaksanaan HSE di Perusahaan dalam Wilayah Yogyakarta* (Yogyakarta: Sosiohumanika)

[5] Nielsen, K J 2013. *Improving Safety Culture Through The Health and Safety Organization: A Case Study*. Vol 48 2014 (Denmark: Elsevier) p7-17.

[6] Dewi, Rijuna 2006. *Pengaruh Keselamatan dan Kesehatan Kerja Terhadap Kinerja Karyawan di PT. Ecogreen Oleoshemicals.* Medan Plant. Unpublished Thesis.

[7] Law No. 1 of 1970 on *Occupational Health and Safety* (Idn).

[8] The Regulation of the Republic of Indonesian Government No. 50 of 2012 *on the Implementation of the Occupational Health and Safety Management System* (OHSMS)

[9] Triatmidi, B 2010. *Kontribusi Pemahaman dan Sikap Guru tentang K3 terhadap Pelaksanaan K3 pada pembelajaran praktik di bengkel Teknik Mekanik Otomotif se-Kota Malang* Vol 33 No. 1; Februari 2010 (Malang: Jurnal Teknologi dan Kejuruan) p 43-54

[10] The Regulation of the Minister of Manpower and Transmigration of the Republic of Indonesia No.PER.15/MEN/VIII/2008 *on the Occupational First Aid* (OFA).

[11] The Regulation of the Indonesian Minister of public Work No. 05/Men/1996 on *Occupational Health and Safety Management System* (OHSMS)

[12] The Regulation of Indonesian Minister of National Education No. 40 of 2008 *on the Standards for Facilities and Infrastructure in Vocational High Schools/Islamic Vocational High Schools.*

[13] The Regulation of the Minister of Public Works No. 09/PER/M/2008 *on the Occupational Health and Safety Management System (OHSMS) Guidelines and Construction of General Jobs* .

[14] Putri, W D 2016 January 19 *Manajemen K3 di SMK akan Diperkuat.* https://www.republika.co.id/berita/nasional/umum/16/01/19/o15xca359-manajemen-k3-di-smk-akan-diperkuat on 28 July 2018.

IOP Conf. Series: Materials Science and Engineering **535** (2019) 012029 doi:10.1088/1757-899X/535/1/012029

Developing learning media for power plant course

F Surwi[1], N Yuniarti[2], and Zamtinah[3]

[1,2,3]Department of Electrical Engineering Education, Faculty of Engineering, Yogyakarta State University, Yogyakarta, Indonesia

E-mail: faranita@uny.ac.id

Abstract. This study aims to (1) develop an appropriate learning media for Power Plant Course, and (2) assess the feasibility of the developed learning media. This is a research and development study using the ADDIE model (Analysis, Design, Development, Implementation, and Evaluation). The research object was the Power Plant Course and the subjects were 30 students of Electrical Engineering Education Department, Faculty of Engineering, Yogyakarta State University. The data were collected by means of questionnaires on a 4-point Likert-type scale. This instrument was validated through expert judgement with the Delphi technique. The obtained data were then analyzed using qualitative descriptive technique. The final product of this research and development study was a learning media used in the Power Plant Course. The results of the feasibility test by the material and media experts as well as by the end users indicate that the developed media is "highly feasible" as the obtained mean scores are 3.41, 3.53, and 3.62 respectively.

1. Introduction

The Department of Electrical Engineering Education (DEEE) employs the Competency Based Curriculum (KBK), i.e. a curriculum of which the learning development, implementation, and evaluation are adjusted to the development of competencies required by the real world of work. Through the implementation of this curriculum, the graduates are expected to have the suitable competencies as required by the industries. Since the 2015/2016 academic year, DEEE has implemented the 2014 Curriculum. The teaching and learning process in the Power Plant Course (PPC) develops the students' competencies in all three domains: cognitive, affective, and psychomotor. PPC covers all the topics related to the operation, maintenance, reparation, and simulation of Steam Power Plant (PLTU), Hydroelectric Power Plant (PLTA), Gas Power Plant (PLTG), Nuclear Power Plant (PLTN), Solar Power Plant (PLTS), Wind Power Plant (PLT Angin), Diesel Power Plant (PLTD), micro-hydro, and generator set. In addition, the students will also deal with topics related to alternator characteristics and generator paralleling. Further, they will have opportunities to have a simulation of Load Frequency Control in a power plant center using the LQR and Robbust methods and conduct field study on a power plant station. The teaching and learning process is carried out in varied activities such as observation, practice, and demonstration, either individually or in-groups. The students are assessed authentically in some aspects on the bases of competency-based assessment, including attendance, attitudes (piety, participation, neatness, cooperation, and work safety), as well as

IOP Conf. Series: Materials Science and Engineering **535** (2019) 012029 doi:10.1088/1757-899X/535/1/012029

knowledge and skills reflected in group practice assignments, individual practice reports, and individual competency/practice tests.

In the implementation of the PPC teaching and learning process in the 2015/2016 academic year, most lecturers still encountered difficulties in achieving the students' targeted competencies. This problem was clearly reflected in the students' PPC scores that were not satisfactory enough. One of the probable triggers of this constraint was the unavailability of supporting learning media in PPC. Besides, the results of an observation in the classroom showed that some students faced difficulties in understanding the materials, did not actively participate during the process, and had low motivation in learning. This condition can be used for evaluation purposes for the lecturers to analyze the causes of the matter and to look for the possible solutions. When viewed from the lecturers' side, these problems are in relation to their teaching styles, strategies, models, media, feedbacks, assignments, assessments, and evaluations. Another constraint in PPC teaching and learning process in the 2015/2016 academic year was the limited availability of learning media, particularly computer-based interactive media. In an effort to overcome those problems, the researchers have developed a custom learning media to help the PPC students in their learning process.

2. Method

This study is a research and development aimed at developing a learning media for the Power Plant Course (PPC) using the Adobe Flash Cs6.

In developing the desired media, the researchers employed the ADDIE model (Analysis, Design, Development, Implementation, and Evaluation) proposed by Branch (2009) [1]. In the Analysis phase, a need analysis for developing the PCC interactive learning media was carried out. This analysis covered three aspects: curriculum, materials, and the students' characteristics. The second phase, Design, dealt with the media presentation aspects, including the navigation structure, interface design, and storyboard. The grand design was then developed in the Development phase through three main activities, namely media development, expert validation, and evaluation. The developed media then was implemented in the fourth phase, Implementation. The last phase, Evaluation, was then carried out in the teaching and learning process of PPC in the theory class.

The research object was the Power Plant Course and the try-out subjects were 30 students from Class A and D, the Department of Electrical Engineering Education, Faculty of Engineering, Yogyakarta State University. The try-out was conducted in the even semester of the 2016/2017 academic year. The data, both quantitative and qualitative, were collected through questionnaires, observations, and interviews. The respondents' answers were rated using the 4-point Likert-type scale, ranging from "less feasible" to "highly feasible". The instruments were validated through expert judgement, whereas the reliabilities were tested using the Cronbach's alpha.

3. Result and Discussion

This research and development study has produced an interactive learning media for the Power Plant Course (PPC). This section will present the details of the developed media.

The very first page is the Introduction Page. Figure 1 shows the Introduction Page.

Figure 1. The Introduction Page

The Home Page is the main page in this PPC interactive media. The user interface of this this page is shown in Figure 2.

Figure 2. The Home Page

The next page is the Material Page that contains all the materials covered in this interactive media.

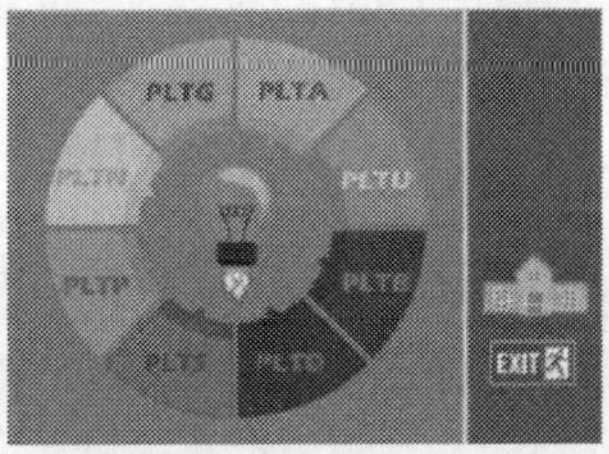

Figure 3. The Material Page

This page shows the Material Page of sub-sections 1, 2, 3, and 4. An example of the Material Page for sub-section 1 is presented in Figure 4.

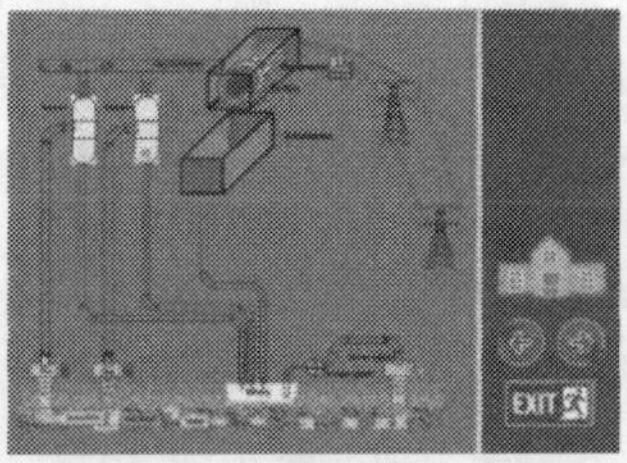

Figure 4. The Material Page of Sub-Section 1

The last page is the Evaluation Page that provides some tasks for the students regarding the presented materials. This evaluation is meant to help the students measure their understanding of the matters.

The developed media is validated by a material expert and a media expert to gain feedbacks. Both experts have validated the media to be feasible to be used in the teaching and learning process in PPC.

3.1 The Feasibility of the Interactive Learning Media of the Power Plant Course

The validation results of the developed media by both experts are presented in Table 1 and 2.

Table 1. The Results of the Expert Judgement towards the Media Quality

Aspects	Mean Scores
Programming (Technical Quality)	3.62
Media Interface (Media Presentation)	3.44
Total	3.53

For the media expert, the developed media is "highly feasible" as it obtains a mean score of 3.53.

Table 2. The Results of the Expert Judgement towards the Material Quality

Aspects	Mean Scores
Instructional Quality	3.30
Content Quality	3.52
Total	3.41

The mean score for the material quality of the developed interactive media is 3.41, indicating that the media is categorized as "highly feasible".

3.2 The Students' Assessment of the Interactive Learning Media of the Power Plant Course

A small and a large group try-out were conducted to obtain data on the students' assessments of the developed interactive media. The questionnaires distributed to the students as the media users covered four aspects: programming (technical quality), interface, materials, and utility. Table 3 presents the mean scores for each of these four aspects.

Table 3. The Students' Assessment of the Interactive Learning Media

Aspects	Mean Scores
Programming	3.55
Interface	3.44
Material	3.67
Utility	38.7
Mean Score	3.62

For the students as the media users, the interactive media is "highly feasible" as it obtains a mean score of 3.62. Thus, it can be said that the developed interactive media for the Power Plant Course is highly feasible to be used in the teaching and learning process.

4. Conclusion

Based on the research results and the discussion on the development of interactive learning media for the Power Plant Course, the researchers have come to the following conclusions: (1) the developed media is categorized as "highly feasible" to be implemented with a mean score of 3.41, viewed from the aspect of the materials presented in it; (2) from the interface and programming aspects, the interactive media is "highly feasible" as well for the obtained mean score is 3.53; and (3) the end users rate this media as "highly feasible" as the mean score for programming, interface, material, and utility aspects is 3.62.

5. References

[1] Branch, R. M., 2009. *Instructional Design: The ADDIE Approach.Department od Educational Psycholology and Instructional Technology University of Georgia*. USA: Springer
[2] Djemari Mardapi. 2008. *Teknik penyusunan tes dan non tes.* Yogyakarta: Mitra Cendekia Press.
[3] Sudarsono. 2013. *Metode penelitian pendidikan. Yogyakarta: UNY PressTiwan (2010) Penerapan modul pembelajaran bahan teknik sebagai upaya peningkatan proses pembelajaran di Jurusan Pendidikan Teknik Mesin FT UNY*. JPTK Vol. 19, No.2. Halmn. 257

IOP Conf. Series: Materials Science and Engineering **535** (2019) 012030 doi:10.1088/1757-899X/535/1/012030

Developing a learning video of the total station set up to improve the students' competencies in geomatics practicum

S Rochmadi[1]

[1]Department of Civil and Planning Engineering, Faculty of Engineering, Yogyakarta State University, Karangmalang, Yogyakarta, Indonesia

E-mail: sunarroch@uny.ac.id

Abstract. This study aimed to develop a learning video of the total station set up, trying out, and evaluating the feasibility of it. This study applied a Define, Design, Develop and Disseminate (4D) research and development model by Thiagarajan. The findings are as follows. (1) The learning video, developed through this study, consists of five main components including opening page, introduction, table of contents, materials, and closing page. (2) The dissemination of the developed media were conducted by implementing the video in Geomatics Practicum. (3) The feasibility of the developed learning video, using 1-4 scale, were scored 3.62 by the learning material expert and 3.44 by the learning media expert. However, the score by the users in Geomatics Practicum was only 3.14.

1. Introduction

Surveying and mapping works, the main topics of Geomatics, are carried out to obtain spatial data and information which are generally presented in maps. These works include three stages, namely data collection, data processing, and data display. Based on the technology development for the terrestrial data collection, the use of optical-conventional devices has shifted to digital electronics such as total station. Total station is a combination of the angle and distance-measuring equipment.

The competency demands related to the total station use should be met, so vocational education remains relevant to the industrial world. However, the topic on conventional-optical device use cannot be easily removed in Geomatics Practicum. It is due to an easier adaptation to move from conventional-optical to digital-electronic measuring equipment, compared to the other way around. Moreover, some schools or industries still use conventional-optical measuring equipment.

Regarding the allocated time for practicum, a dilemma has appeared. If the practicum topic on total station use is added, it will significantly reduce the time to master conventional-optical equipment. Taking into account this, it is necessary to accommodate the additional practicum topic about the total station use. Therefore, it is urgent to enhance the practicum topics without time allocated addition, by exploring more effective learning media.

Audio visual media in the form of learning videos can be used to accommodate these needs. On the internet nowadays, like on *youtube.com*, there have been many videos about the total station set up. However, the existing videos have not completely applied the learning standard. For example, it is not clear about what the learning objectives are, who the learners are, how the learning preconditions are, and how the learning implementation and evaluation are. Furthermore, the manual of the total station equipment, which can be downloaded, cannot be easily understood by students.

To clearly show the work stages in the vocational learning, a video can be employed [1]. Animation in videos can provide illustrations of fast and invisible processes. For example, due to a safety reason, the process is covered by a case. Information obtained by the students through knowledge construction and understanding will be stored in the long-term memory [2].

As mentioned in the competencies in Geomatics Practicum, students should be able to carry out surveying and mapping works. In addition, the students of Civil Engineering and Planning Education Department should master the competencies of surveying and mapping to plan the construction works, and to teach the construction vocational students.

A screencast-based spreadsheet learning media has been developed [3]. The development is carried out in the following stages: (1) preparation (the devices and materials needed are prepared); (2) recording (all activities carried out on the monitor screen are recorded); (3) editing (text, audio or narration, zoom and animation effects are added); (4) publishing (the video is created into one unit and converted into mp4 format, and (5) finishing (quizzes are made and combined with the video).

A multimedia-based interactive media in the Basic Electricity course has been developed [4]. The development stages are: (1) conducting a needs analysis, (2) designing, (3) translating the design module into an application, (4) testing the learning media software, (5) trying out the product, and (6) evaluating the learning media based on the user' input.

The learning media can be used to convey knowledge and skills. The video can deliver the learning materials, so the learning approach shifts from teacher-centered to learner-centered, the teacher as a facilitator and students as the actors in the learning process.

This study aims to: (1) develop a learning video of the total station set up to improve the students' competencies in Geomatics Practicum, (2) try out, and (3) evaluate the feasibility of the product.

2. Method

This study employed four main stages, namely defining, designing, developing, and disseminating (4D) research and development model according to Thiagarajan [5]. The subjects of this study were the students in the Civil Engineering and Planning Education Department, Faculty of Engineering, Yogyakarta State University, taking the Geomatics Practicum II course. The objective of this study was to develop a learning video of the total station set up. The data were collected to evaluate the feasibility of the developed media. Furthermore, the data were collected by observation, questionnaires, and interviews.

The data were analyzed qualitatively and quantitatively. A qualitative descriptive analysis was applied to the observation and interview data. The questionnaire data were analyzed using a quantitative descriptive analysis. The results of the analysis were used as the input to revise the developed learning video.

The quantitative analysis was applied on the questionnaire data. The data were in the assessment scale forms consisting of four scales, namely feasible, quite feasible, less feasible, and not feasible. The data analysis was in the following stages: (1) data tabulation, obtained from the media assessment items from each expert, and (2) calculation on the average total score of each assessment aspect. The questionnaires used the options of strongly agree (4), agree (3), disagree (2), and strongly disagree.

3. Result and Discussions

The developed product is a learning video of 724,902 kilobyte (KB) in an MP4 format with the duration of 10 minutes 6 seconds. The video can be accessed in https://www.youtube.com/watch?v=UxWfoxRskDM. The contents of the learning media emphasize the introduction of equipment used in the total station such as tripod, reflector prism, thumbtack, and umbrella. Furthermore, the total station set up is explained, as well as how to set up the tripod, set the total station on the tripod, get the total station ready to measure, and read the measurement.

The product of the designed and systematic media development includes the following components: (1) opening page, containing the tittle, design team, competencies, and objectives; (2) introduction, containing a description of the competencies and objectives of the learning media; (3) video materials, containing an explanation of the total station use, starting from determining the measuring point, setting the instrument on the tripod, centering the total station to the measuring point, setting the total station to be ready to measure, and reading the measurement; and (4) closing page, containing the acknowledgement to the user.

The validation by the learning material expert is based on the three aspects, namely learning objective, material delivery, and motivation quality. With a scale of 1-4, eight indicators obtain a score of 4, and five indicators achieve a score of 3. Overall, the feasibility score by the learning material expert is 3.62. However, before implemented, the learning media needs to be evaluated and improved based on the material expert's advice.

The validation was also conducted by the learning media expert in terms of three aspects, namely software engineering, learning design, and visual communication. With a scale of 1-4, twelve indicators obtain a score of 4, and fifteen indicators achieve a score of 3. Overall, the feasibility score given by the learning media expert was 3.44. However, before implemented, the learning media needs to be evaluated and improved based on the media expert's input.

The data collection regarding the product feasibility analysis used questionnaires filled by the users. The users were the students participating in the Geomatics Practicum II, after watching the developed learning video. There were four aspects assessed, namely generality, material, appearance, and learning.

The score of the overall feasibility assessment by the users is 3.14 on a scale of 1-4. The three indicators scoring less than 3.00 are presented on Table 1.

Table 1. Indicators scoring less than 3.00 by the users.

	Indicators	Average Scores
1	The material presented in the learning video attracts the students' interest to learn further materials.	2.47
2	The material presented in the learning video is easy to understand.	2.88
3	The material scope presented in the learning video is sufficient for the students to learn the next material.	2.88

The feasibility scores of the learning video, in the 1-4 rating scale, are 3.62 by the material expert and 3.44 by the media expert. However, the score by the users only 3.14. There are even three indicators scored less than 3.00.

The inputs from the users can be categorized into two groups, namely the sounds that need to be clarified and harmonized, and the explanations that need to be made more detailed and easier to understand. In addition, the users' responses to this learning media showed that they agree to use the learning video. Nevertheless, regarding the low score by the users, the learning video cannot replace the explanation and demonstration by the lecturer, but it complement and assist the learning process. It is said that the learning video can only be used as the supplementary material, because it cannot substitute the explanation and demonstration by the lecturer.

4. Conclusion

The conclusions of this study are as follows. (1) The learning video of the total station use to improve the students' competencies in Geomatics Practicum has been developed. (2) The learning media has five main components such as opening page, introduction, table of content, video materials, and closing page. (3) The feasibility scores of the developed learning video, on a scale of 1-4, are as follows. (a) The score by the learning media expert is 3.44, based on the aspects of software engineering, learning design, and visual communication. (b) The score by the learning material expert is 3.62, based on the aspects of learning objectives, material delivery, and motivation quality. (c) The score by the users is 3.14, based on all of the aspects. In addition, the students' responses to this learning media show that they agree to use the video. However, the use of the learning video cannot completely replace the explanation and demonstration by the lecturer, but it can be the supplementary material in their learning process.

5. References

[1] Howe F and Staden C 2015 Work process oriented and multimedia-based learning in vocational education and training ed Gessler M and Freud L *Crossing boundaries in vocational education and training: Innovative concepts for the 21st century, Conference Proceeding, Institute Technology and Education* (Bremen: University of Bremen) p 25-32

[2] Santrock J W 2008 *Psikologi Pendidikan* Edisi kedua terjemahan Tri Wibowo B S (Jakarta: Kencana Prenada Media Group)

[3] Munir M 2013 Analisis pengembangan media pembelajaran pengolah angka (spreadsheet) berbasis video screencast *Jurnal Pendidikan Teknologi dan Kejuruan* 21 (4) p 307-313

[4] Mustholiq I, Sukir and Chandra A 2007 Pengembangan media pembelajaran interaktif berbasis multimedia pada mata kuliah Dasar Listrik *Jurnal Pendidikan Teknologi dan Kejuruan* 16 (1) p 1-18

[5] Thiagarajan S, Semmel D D and Semmelpp M I 1974 *Instructional Development for Training Teachers of Exceptional Children* (Minnesotta: U.S. Office of Education)

Institute of Physics
Dirac House, Temple Back
Bristol BS1 6BE UK

ISSN: 1757-8981
ISBN 978-1-5108-8921-7
9 781510 889217